The Craft of Plant Cultivation

MW01675425

Techniques for Growing and Cultivating Flora

Aubrey Munoz

The Craft of Plant Cultivation

Chapter 1: The Fundamentals of Plant Propagation

Understanding Plant Propagation Basics

Plant propagation is an ancient practice that has allowed humanity to cultivate and sustain a diverse array of plant species. This art, grounded in both science and tradition, serves as the backbone of horticulture, agriculture, and even conservation efforts. At its core, plant propagation is the process of creating new plants from a variety of sources, including seeds, cuttings, and other plant parts. Understanding the basics of this practice is crucial for anyone looking to delve into the world of gardening or professional horticulture.

The journey of plant propagation begins with recognizing the life cycle of plants. Every plant starts its life from a seed, which is a small, dormant entity packed with potential. A seed waits patiently for the right conditions—moisture, warmth, and light—to germinate and spring into life. This transformation from seed to seedling is one of the most fascinating phases of plant propagation, as it involves a series of complex biochemical processes. The energy stored within the seed is utilized in these early stages, driving the growth of roots and shoots.

However, seeds are just one part of the propagation puzzle. Vegetative propagation, another fundamental concept, involves using parts of an existing plant to create new ones. This method capitalizes on the plant's ability to regenerate and grow, making it a popular choice for many gardeners and horticulturists. Techniques such as cuttings, layering, and division fall under this category, each with its own unique approach and benefits.

Key to successful propagation is understanding the conditions required by different plant species. Each species has its own specific needs in terms of light, temperature, humidity, and soil composition. Recognizing these needs and replicating them in a controlled environment can significantly increase the chances of successful propagation. This is where the role of a gardener or horticulturist becomes part scientist, part artist—balancing the precise requirements of each species with the creative intuition that comes from experience.

Terminology plays a crucial role in demystifying the world of plant propagation. Concepts such as "stratification" and "scarification" refer to specific techniques used to break seed dormancy and encourage germination. Stratification involves exposing seeds to a period of cold, mimicking winter conditions to trigger germination when warmth returns. Scarification, on the other hand, entails physically breaking or softening the seed coat to allow moisture to penetrate, facilitating germination.

Understanding these terms and techniques can empower gardeners to tackle a wide range of propagation challenges. It also fosters a deeper appreciation for the intricate dance between nature and nurture that defines successful plant cultivation.

The role of propagation in horticulture extends beyond simply increasing plant numbers. It is a vital tool for preserving genetic diversity, developing new plant varieties, and even combating the impacts of climate change. By selecting and propagating plants with desirable traits, horticulturists can create resilient species that thrive in changing environments. This selective breeding, combined with traditional propagation techniques, helps maintain healthy ecosystems and contributes to global food security.

Selecting the right plants for propagation is a critical step that often determines the success of the entire process. Factors such as the plant's growth habits, its adaptability to the local climate, and its potential for disease resistance should all be considered. Additionally, the intended use of the propagated plants—whether for ornamental purposes, food production, or habitat restoration—can influence the choice of species and propagation method.

For the novice gardener, understanding these basics is an invitation to experiment and explore. The world of plant propagation is rich with possibilities, offering endless opportunities to cultivate beauty, sustenance, and sustainability. Each new plant grown from seed or cutting represents a triumph of life, an affirmation of the gardener's skill and patience.

To truly grasp the fundamentals, it is essential to engage in hands-on practice. Observing how different seeds respond to various germination techniques or how cuttings root in different soil mediums provides invaluable insights. This experiential learning is complemented by a wealth of resources available to modern gardeners, from books and online forums to local gardening clubs and workshops.

Moreover, the historical context of plant propagation offers a fascinating window into human ingenuity. Throughout history, civilizations have developed and refined propagation techniques to suit their unique environmental conditions and cultural needs. From the ancient Egyptians' use of cuttings to cultivate grapevines to contemporary tissue culture methods that allow for mass production of orchids, the evolution of propagation reflects our enduring relationship with the plant kingdom.

In recent years, advancements in technology have opened up new frontiers in plant propagation. Innovative techniques such as micropropagation and hydroponics are transforming the way we think about growing plants, offering solutions to some of the most pressing challenges in agriculture and conservation. These methods, which will be explored in later chapters, are expanding the possibilities of what can be achieved through thoughtful and informed propagation practices.

Ultimately, understanding the basics of plant propagation is about unlocking the potential within each seed, each cutting, and each budding plant. It is a reminder of the remarkable resilience and adaptability of nature, and of our ability to harness these qualities to create a more verdant and sustainable world. By delving into this foundational knowledge, gardeners and horticulturists alike can cultivate not only plants but a deeper connection to the natural world and its endless cycles of growth and renewal.

Historical Context and Importance

Throughout the annals of human history, the cultivation and propagation of plants have been intertwined with the very survival and advancement of civilizations. From the early days of agriculture to the sophisticated horticultural practices of today, the ability to propagate plants has not only sustained human populations but also facilitated cultural and technological evolution. Understanding the historical context of plant propagation provides insight into its profound importance, shaping societies and ecosystems across the globe.

Tracing back to the dawn of agriculture, plant propagation emerged as a transformative force that allowed humans to transition from nomadic lifestyles to settled agricultural communities. This shift, often referred to as the Neolithic

Revolution, marked a pivotal moment in human history, as people learned to cultivate grains, fruits, and vegetables. The domestication of wild plants was not a simple endeavor; it required the observation and experimentation with different propagation techniques, ultimately leading to the selection of varieties that were more productive and easier to cultivate.

In ancient Egypt, for instance, the Nile River's fertile banks became the cradle of plant propagation practices that would influence agriculture for centuries. Egyptians meticulously documented their horticultural methods, including the use of cuttings to propagate figs and grapes. These techniques were depicted in tomb paintings and agricultural texts, highlighting the importance of plant propagation in their society. The ability to reliably grow food and medicinal plants contributed to Egypt's prosperity and played a crucial role in its lasting legacy.

Similarly, in ancient China, plant propagation was elevated to an art form, with sophisticated methods developed for cultivating rice, tea, and a variety of fruits. The Chinese were pioneers in the development of grafting techniques, which allowed them to improve fruit quality and increase yields. This innovation not only enhanced food production but also laid the groundwork for botanical studies and the systematic classification of plants.

Across the Mediterranean, the Greeks and Romans expanded upon earlier propagation practices, integrating them into a burgeoning agricultural economy. The Greeks, in particular, were known for their botanical gardens, where they studied and experimented with plant cultivation techniques. These gardens served as centers of learning and exploration, where scholars like Theophrastus documented their findings, contributing to the foundation of botany as a scientific discipline.

The Roman Empire, with its vast territories and diverse climates, relied heavily on plant propagation to support its extensive agricultural network. Romans implemented advanced irrigation systems and crop rotation practices to maximize productivity. The introduction of new plant species from conquered lands led to the expansion of horticultural knowledge and the diversification of agricultural practices. Roman innovations in propagation techniques, such as air layering and division, were instrumental in the Empire's ability to sustain large urban populations.

As the Middle Ages unfolded, monastic gardens became vital centers for the preservation and propagation of plant species. Monks meticulously cultivated medicinal herbs and food crops, utilizing propagation methods passed down through generations. These gardens were not only sources of sustenance but also repositories of botanical knowledge during a time when much of Europe experienced intellectual stagnation.

The Age of Exploration ushered in a new era of plant propagation, as explorers and botanists traversed the globe, collecting plant specimens and seeds from distant lands. This period of discovery led to the introduction of countless new species to Europe, Asia, and the Americas, challenging horticulturists to adapt and refine their propagation techniques. The exchange of plants and knowledge between continents enriched global biodiversity and expanded the agricultural capabilities of societies worldwide.

In the 18th and 19th centuries, the Industrial Revolution brought about significant advancements in plant propagation. The development of glasshouses and controlled environments allowed for year-round cultivation and experimentation. This period saw the rise of botanical gardens as centers of research and education, where scientists could study plant growth and reproduction under controlled conditions. The propagation of ornamental plants became increasingly popular, leading to the establishment of commercial nurseries and the widespread dissemination of horticultural knowledge.

The 20th century introduced groundbreaking scientific discoveries that revolutionized plant propagation. The understanding of genetics and plant physiology provided a framework for the development of new propagation techniques, such as tissue culture and hybridization. These innovations enabled the mass production of plants with desirable traits, such as disease resistance and increased yield, transforming agriculture and horticulture.

Today, plant propagation continues to be a cornerstone of food security, ecological restoration, and conservation efforts. As the global population grows and environmental challenges intensify, the ability to propagate plants efficiently and sustainably is more important than ever. Modern propagation techniques, supported by advances in biotechnology and environmental science, offer solutions to some of the most pressing issues facing humanity, from climate change to habitat loss.

The historical journey of plant propagation underscores its enduring importance as both a practical and scientific endeavor. It is a testament to human ingenuity and adaptability, reflecting our deep connection to the natural world and our ability to harness its potential for the betterment of society. As we continue to explore and innovate in the realm of plant propagation, we honor the legacy of those who came before us, while paving the way for future generations to thrive in harmony with the environment.

Key Terminology and Concepts

Embarking on the journey of plant propagation requires a solid grasp of the terminology and concepts that underpin this intricate art. Not only do these terms serve as a common language among horticulturists, but they also provide the foundation for understanding the processes involved in creating new plant life. As you navigate the world of plant propagation, familiarizing yourself with these key terms and concepts will empower you to make informed decisions and cultivate a thriving garden.

Propagation itself is the act of reproducing plants, encompassing a range of techniques from seed sowing to cloning. It's a concept that bridges biology with practical horticulture, grounded in the understanding of how plants grow and reproduce. Central to this concept is the idea of plant genetics and heredity. Each plant carries a unique genetic blueprint that determines its traits, from flower color to growth habit. When propagating plants, especially through seeds, these genetic traits are passed on to the next generation, though variations can occur due to genetic recombination.

A fundamental term in plant propagation is "germination," which refers to the process by which a seed develops into a new plant. Germination begins when a seed absorbs water, triggering metabolic processes that awaken the dormant embryo within. This leads to the emergence of the radicle, or embryonic root, followed by the shoot. Successful germination depends on a combination of factors, including temperature, moisture, and light conditions. Understanding these requirements is essential for ensuring that seeds sprout into healthy seedlings.

For some seeds, achieving germination requires overcoming dormancy, a survival mechanism that prevents seeds from germinating until conditions are favorable. Two common methods to break dormancy are stratification and scarification. Stratification involves exposing seeds to a period of cold temperatures, simulating winter conditions, which primes them for germination when temperatures rise. Scarification, on the other hand, involves physically altering the seed coat through abrasion or other means to allow water penetration, thus promoting germination.

Vegetative propagation is another key concept, referring to the process of creating new plants from the vegetative parts of a parent plant, such as stems, leaves, or roots. This method bypasses the seed stage entirely, relying on the plant's ability to regenerate from existing tissues. One popular technique under this category is "cuttings," where a section of the plant is cut and encouraged to develop roots, ultimately forming a new plant. Cuttings can be taken from different parts of the plant, including stem cuttings, leaf cuttings, and root cuttings, each offering unique advantages depending on the plant species.

"Layering" is another vegetative propagation technique, involving the rooting of a plant's branch while it is still attached to the parent plant. This method takes advantage of a plant's natural tendency to develop roots when in contact with the soil. By bending a branch down and covering it with soil, the buried section can produce roots, allowing it to be severed from the parent plant once established. Layering is particularly useful for plants that are difficult to propagate through cuttings.

The terms "grafting" and "budding" are vital to understanding advanced propagation techniques. Grafting involves joining the tissues of two plants so that they grow as one. Typically, the upper part, called the scion, is attached to the rootstock of another plant. This method is commonly used in fruit tree propagation to combine desirable traits from two different plants, such as disease resistance in the rootstock with superior fruit quality in the scion. Budding is a similar process, where a single bud from the desired plant is inserted into the rootstock.

Another critical concept in plant propagation is the role of hormones, specifically "auxins." Auxins are plant hormones that regulate growth and play a significant role in rooting processes. When propagating plants through cuttings, the application of rooting hormones, which contain synthetic auxins, can enhance root development and increase the chances of successful propagation. Understanding how these hormones work and their appropriate application can significantly influence the outcome of vegetative propagation efforts.

"Hardening off" is a term often encountered in propagation, especially when transitioning plants from controlled environments to outdoor conditions. This process involves gradually acclimating plants to the external environment, reducing the risk of shock or stress. By slowly exposing plants to outdoor conditions, such as varying temperatures and sunlight levels, they develop resilience and adaptability, ensuring a smoother transition and better survival rates.

The concept of "micropropagation" represents a modern advancement in plant propagation, involving the use of tissue culture techniques to produce large numbers of identical plants from a small tissue sample. This method is particularly valuable for mass-producing plants with specific traits or for conserving rare and endangered species. Micropropagation relies on sterile laboratory conditions and precise control of growth factors, allowing for rapid multiplication and consistent quality.

In understanding these key terms and concepts, it's important to recognize the interconnectedness of plant propagation with broader ecological and agricultural systems. Propagation techniques have evolved alongside human knowledge and technology, adapting to meet the needs of diverse environments and changing climates. The ability to propagate plants successfully contributes to biodiversity, food security, and the preservation of genetic resources.

As you embrace the world of plant propagation, these foundational terms and concepts serve as your guide, illuminating the path to successful cultivation. Whether you're sowing seeds for a vegetable garden, grafting fruit trees for a bountiful orchard, or experimenting with cutting techniques for ornamental plants, a solid understanding of these principles will enhance your skills and deepen your appreciation for the art and science of plant propagation. With this knowledge, you are equipped to explore the endless possibilities that plant propagation offers, cultivating not only plants but also a deeper connection to the natural world.

The Role of Propagation in Horticulture

Horticulture, the art and science of cultivating plants, thrives on the principles and practices of propagation. As the backbone of horticultural endeavors, propagation serves a multifaceted role, from ensuring the sustainability of plant species to enhancing the aesthetic and nutritional value of our environments. Its significance reaches far beyond the simple act of multiplying plants, intertwining with ecological preservation, economic development, and cultural traditions.

The essence of propagation in horticulture lies in its ability to produce plants on a large scale, ensuring a steady supply for various needs. This is particularly crucial for ornamental horticulture, where demand for flowers, shrubs, and decorative plants is ever-growing. Nurseries and garden centers depend on efficient propagation techniques to meet consumer demands, offering a diverse array of plant species and cultivars. Through techniques such as seed sowing, cuttings, and grafting, horticulturists can replicate plants that exhibit desirable traits, such as vibrant colors, unique shapes, or resilience to pests and diseases.

Propagation also plays a pivotal role in food production, a cornerstone of horticulture. Fruits, vegetables, and herbs are propagated to maintain and increase agricultural output, supporting both small-scale farmers and large commercial operations. The ability to propagate high-yielding and disease-resistant varieties is essential for ensuring food security and adapting to changing environmental conditions. This becomes increasingly important as the global population rises and climate change affects traditional growing regions.

Beyond its practical applications, propagation is a key player in biodiversity conservation. By enabling the cultivation of native and rare species, it helps maintain genetic diversity within ecosystems. Botanical gardens and conservation organizations often use propagation to preserve endangered plant species, ensuring their survival for future generations. This not only protects the plants themselves but also supports the myriad of organisms that rely on them, fostering balanced ecosystems.

Propagation techniques are instrumental in habitat restoration projects, where they assist in reestablishing native vegetation in degraded areas. By carefully selecting and propagating native plant species, horticulturists can restore ecological balance, improve soil health, and increase biodiversity. Such efforts are vital in counteracting the impacts of deforestation, urbanization, and agricultural expansion, providing a pathway for ecological recovery and resilience.

In the realm of landscape design, propagation opens the door to creativity and innovation. Landscape architects and gardeners leverage propagation to experiment with plant combinations, textures, and colors, crafting spaces that are both functional and aesthetically pleasing. The availability of a wide range of propagated plants allows for the creation of gardens that reflect personal style, cultural heritage, or the natural beauty of a region.

For the home gardener, propagation offers an accessible entry point into the world of horticulture. Whether propagating houseplants, cultivating a vegetable garden, or experimenting with exotic blooms, the practice provides a sense of accomplishment and connection to nature. It empowers individuals to nurture life, transform spaces, and contribute to environmental sustainability, no matter the scale.

Propagation's role extends to education and research, forming the foundation for botanical studies and horticultural training programs. Through hands-on experience with propagation techniques, students and researchers gain insights into plant biology, genetics, and ecology. This knowledge fuels innovation, leading to the development of new cultivars, sustainable practices, and solutions to emerging challenges in horticulture.

Economic benefits also stem from the role of propagation in horticulture. The production and sale of propagated plants generate significant revenue, supporting local economies and creating employment opportunities.

Nurseries, garden centers, and landscaping businesses thrive on the propagation of plants, contributing to the economic vitality of communities. Additionally, the export of propagated plants and plant materials can bolster international trade, connecting regions and cultures through a shared appreciation for horticulture.

As technological advancements continue to shape the horticultural landscape, propagation remains at the forefront of innovation. Techniques such as tissue culture and genetic modification enhance the efficiency and precision of propagation, enabling the rapid multiplication of plants with specific traits. These advancements hold promise for addressing global challenges, such as food scarcity and environmental degradation, by developing plants that are better adapted to future conditions.

The cultural significance of propagation cannot be overlooked, as it is deeply woven into the fabric of societies around the world. Many cultures have long-standing traditions of plant cultivation and propagation, which are passed down through generations. These practices are often imbued with symbolism and meaning, reflecting the values, beliefs, and histories of communities. From the bonsai art of Japan to the heirloom gardens of Europe, propagation serves as a bridge between the past and the present, preserving cultural heritage and fostering a sense of identity.

Propagation's role in horticulture is as dynamic as it is essential, encompassing a spectrum of activities that support the health and vitality of our planet. It is a testament to the intricate relationship between humans and plants, a relationship that is constantly evolving and adapting to meet the challenges and opportunities of the modern world. By understanding and harnessing the potential of propagation, we can cultivate a future that is rich in biodiversity, sustainability, and beauty. Whether on a global scale or within the confines of a backyard garden, the influence of propagation resonates, nurturing life and inspiring growth.

Selecting the Right Plants for Propagation

Selecting the right plants for propagation is a crucial step that can determine the success of your horticultural endeavors. This selection process is akin to choosing the right ingredients for a recipe; each plant has its own unique characteristics and requirements that must be understood and considered. By carefully choosing plants that suit your goals, environment, and skill level, you can optimize your chances of successful propagation and cultivate a thriving garden.

The first consideration in plant selection is the intended purpose of propagation. Are you planning to grow plants for aesthetic appeal, food production, or perhaps to conserve a particular species? Understanding your primary objective will help narrow down the vast array of plant options. For example, if your goal is to create a vibrant ornamental garden, you might lean towards flowering plants such as roses, hydrangeas, or lilies. Conversely, if food production is your aim, selecting vegetables, fruits, or herbs that thrive in your climate should be a priority.

Climate and environmental conditions play a significant role in plant propagation. Each plant species has specific needs in terms of temperature, humidity, and sunlight. Understanding your local climate is essential to selecting plants that will thrive in your garden. For instance, succulents and cacti are well-suited to arid environments with high temperatures and low humidity, while ferns and mosses prefer cooler, shaded areas with ample moisture. Conducting a thorough assessment of your garden's microclimate will guide you in choosing plants that are naturally adapted to those conditions.

Soil type and quality are also critical factors in plant selection. Different plants have varying soil preferences, ranging from well-draining sandy soils to nutrient-rich loams. Conducting a soil test can provide valuable insights into the composition and pH of your soil, helping you select plants that are compatible with it. For example, blueberries thrive in acidic soils, whereas lavender prefers alkaline conditions. If your soil does not naturally meet the needs of your desired plants, amending it with organic matter or other additives can adjust its properties to better support your choices.

Another important aspect of plant selection is disease resistance and pest tolerance. Some plant varieties are more susceptible to diseases and pests, which can lead to significant challenges in propagation and cultivation. Researching the common issues faced by different plants and choosing varieties known for their resilience can save time and effort in the long run. Many horticulturalists and seed companies offer disease-resistant cultivars that are bred to withstand specific threats, providing a more robust option for propagation.

The growth habit and space requirements of plants should also be considered. Understanding how a plant grows, spreads, and matures will help you plan your garden layout and ensure that each plant has the space it needs to thrive. Vining plants, such as tomatoes or climbing roses, require support structures and ample vertical space, while bushy plants like shrubs or perennials may need more horizontal room to spread. By selecting plants that fit the available space and complement each other, you can create a harmonious and visually appealing garden.

Timing is another crucial factor in selecting plants for propagation. Different plants have varying propagation windows, influenced by their growth cycles and environmental triggers. Some seeds require stratification, a period of cold exposure, before they will germinate, while others may need to be sown immediately after harvest. Understanding these timing requirements will help you plan your propagation activities and ensure that your plants have the best chance of success.

For beginners, starting with easy-to-propagate plants can build confidence and provide valuable learning experiences. Some plants, such as spider plants, pothos, and mint, are known for their vigorous growth and adaptability, making them ideal choices for novice gardeners. These plants often require minimal care and can tolerate a range of environmental conditions, providing a forgiving introduction to the world of propagation.

It's also beneficial to consider the availability of plant materials for propagation. Access to seeds, cuttings, or divisions can vary based on the plant species and your location. Local nurseries, botanical gardens, and plant exchanges are excellent resources for acquiring propagation materials. Additionally, connecting with gardening communities and online forums can provide valuable tips and advice on sourcing and propagating specific plants.

Sustainability and environmental impact are increasingly important considerations in plant selection. Opting for native plants, which are naturally adapted to your region, can enhance local biodiversity and reduce the need for excessive watering, fertilizers, or pesticides. Native plants often support local wildlife, including pollinators and beneficial insects, contributing to a balanced ecosystem. By choosing plants that align with sustainable gardening practices, you can create a garden that is not only beautiful but also environmentally friendly.

Ultimately, selecting the right plants for propagation is a thoughtful process that requires careful consideration of multiple factors. By aligning your choices with your goals, environment, and available resources, you can create a thriving garden that reflects your personal style and values. Whether you're cultivating a small urban oasis or a sprawling rural landscape, the plants you choose will shape the character and success of your horticultural journey, offering endless opportunities for growth and discovery.

Chapter 2: Methods of Seed Propagation

Seed Selection and Collection

Seed selection and collection form the cornerstone of successful plant propagation. The journey from seed to sprout is a delicate one, requiring careful attention to detail and a deep understanding of the plants you wish to cultivate. By mastering the art of choosing and gathering seeds, you set the stage for a flourishing garden, rich with diversity and vitality.

Selecting the right seeds begins with a clear understanding of your gardening goals. Are you aiming to cultivate a vegetable garden, enhance the aesthetic appeal of your landscape, or perhaps nurture rare and endangered species? Defining your purpose will guide your selection process, ensuring that the seeds you choose align with your vision. For instance, if your goal is to grow a vegetable garden, consider the types of produce you and your family enjoy and the climate suitability of those plants. Alternatively, for ornamental purposes, think about the color schemes and plant textures that will complement your existing garden.

Once you have a vision in mind, research becomes your ally. Delve into the characteristics of various plant species and cultivars, paying close attention to their growth habits, pest resistance, and environmental requirements. Seed catalogs, gardening books, and online resources provide a wealth of information to help you make informed decisions. For beginners, starting with easy-to-grow plants such as radishes, sunflowers, or marigolds can be a rewarding experience. These plants are known for their resilience and adaptability, making them ideal choices for novice gardeners.

Quality is paramount when it comes to seed selection. High-quality seeds are the foundation of healthy plants, offering a greater chance of successful germination and robust growth. When purchasing seeds, opt for reputable suppliers who provide detailed information about the seed's origin, viability, and any treatments they may have undergone. Look for seeds that are plump, uniform in size, and free of discoloration or damage. Many seed packets include germination rates and planting instructions, which can be invaluable for planning your propagation activities.

Collecting seeds from your garden or local environment is a cost-effective and sustainable way to expand your plant collection. This practice also allows you to preserve the genetic traits of plants that thrive in your specific conditions. Timing is critical when collecting seeds, as they must be harvested at the right stage of maturity. Immature seeds may not germinate effectively, while overripe seeds risk dispersal or damage. Observing the plant's life cycle and becoming familiar with the visual cues that indicate seed readiness will enhance your collection skills.

Seed collection techniques vary depending on the plant species and seed type. For many flowering plants, seeds can be found in pods or capsules that dry and split open when mature. Gently shaking the seed head over a container or using a small pair of scissors to clip the pods can help you gather seeds without damaging the plant. For fleshy fruits, such as tomatoes or berries, the seeds are often embedded within the fruit and require careful extraction and cleaning. This usually involves scooping out the seeds, rinsing away the pulp, and drying them thoroughly before storage.

Proper seed storage is crucial to maintaining seed viability. Once collected, seeds should be cleaned and dried to prevent mold or decay. Store seeds in a cool, dry, and dark place, such as an airtight container in a refrigerator or a dedicated seed storage box. Label each container with the plant species, collection date, and any relevant notes to keep your collection organized and accessible. Monitoring seed viability over time is important, as germination rates can decline with age. Conducting periodic germination tests can help you assess the quality of your seeds and make informed decisions about planting.

For those interested in preserving heirloom or rare plant varieties, seed saving becomes an essential practice. Heirloom seeds are open-pollinated and have been passed down through generations, often carrying unique traits and flavors. Saving these seeds helps maintain biodiversity and cultural heritage, offering a connection to the past and a resource for the future. When saving heirloom seeds, be mindful of cross-pollination, which can occur when plants of the same species exchange pollen. To preserve the integrity of your heirloom seeds, consider techniques such as isolation or hand-pollination.

Seed exchanges and community seed banks offer valuable opportunities to diversify your seed collection and connect with fellow gardeners. These initiatives promote the sharing of seeds, knowledge, and resources, fostering a sense of community and cooperation. Participating in a seed exchange allows you to access a wider range of plant varieties, learn from experienced gardeners, and contribute to the conservation of plant biodiversity. Whether through local gardening clubs or online forums, engaging with these networks can enrich your gardening experience and inspire new ideas.

As you embark on the journey of seed selection and collection, remember that patience and observation are your greatest allies. Each seed carries the potential for life, a promise of growth and transformation. By choosing seeds with care and collecting them with respect for nature's cycles, you cultivate not only a garden but also a deeper understanding of the interconnectedness of all living things. With each seed planted, you take part in the timeless dance of creation, nurturing a legacy of beauty and abundance for generations to come.

Techniques for Successful Germination

The process of germination marks the miraculous moment when a dormant seed bursts into life, setting the stage for a plant's journey from tiny sprout to full-grown maturity. For gardeners and horticulturists alike, mastering the techniques for successful germination is essential, as this crucial stage determines the future health and vigor of plants. By understanding and applying the right methods, you can maximize germination rates and nurture seeds into robust seedlings ready for transplanting.

The journey of a seed begins with environmental triggers that awaken its dormant state. Water is the primary catalyst, initiating a cascade of biochemical processes that transform the seed from a quiescent state into an active, growing organism. Achieving optimal moisture levels is critical; seeds must absorb enough water to swell and activate their metabolic machinery without becoming waterlogged, which can lead to rot. The key lies in maintaining consistent moisture through gentle watering or the use of a humidity dome to create a controlled environment.

Temperature is another crucial factor influencing germination. Each plant species has its own preferred temperature range for germination, typically within the range of 60 to 80 degrees Fahrenheit. Understanding the specific temperature requirements of your seeds is vital, as temperatures that are too low can delay or inhibit germination, while excessive heat can damage seeds. Many gardeners use a seedling heat mat to provide gentle warmth, ensuring that seeds remain within their optimal temperature range and encouraging faster germination.

Light exposure plays a variable role in the germination process. Some seeds, known as photoblastic seeds, require light to germinate, while others prefer darkness. It's important to know the light requirements of your seeds and adjust your setup accordingly. For instance, light-dependent seeds can be sown on the surface of the soil and exposed to sunlight or grow lights, while those that prefer darkness can be lightly covered with soil. Properly managing light exposure ensures that seeds receive the necessary cues to begin germinating.

Soil or growing medium is the foundation upon which seeds germinate, providing essential support and nutrients. A well-draining, sterile seed-starting mix is ideal for promoting healthy germination, as it reduces the risk of disease and provides the right balance of moisture and aeration. Avoid using garden soil, which may contain pathogens or compact easily, hindering root development. Pre-moisten the growing medium before sowing seeds to ensure even moisture distribution, and lightly tamp down the surface to create good seed-to-soil contact.

Stratification and scarification are techniques used to break seed dormancy, particularly for seeds with hard coats or specific environmental triggers. Stratification involves exposing seeds to a period of cold, moist conditions, mimicking winter and priming them for germination once temperatures rise. This can be achieved by placing seeds in a damp medium and refrigerating them for several weeks. Scarification, on the other hand, involves physically altering the seed coat through abrasion, soaking, or nicking, allowing water to penetrate and trigger germination. Understanding the dormancy-breaking requirements of your seeds can greatly enhance germination success.

Spacing and depth are critical considerations when sowing seeds. Overcrowding can lead to competition for light, water, and nutrients, while sowing seeds too deeply may prevent them from reaching the surface. Each seed packet typically provides guidelines for spacing and depth, which should be followed closely. As a general rule, seeds should be planted at a depth approximately twice their diameter, though this can vary based on seed size and species. Proper spacing and depth ensure that seedlings have room to grow and access the resources they need to thrive.

Once seeds have germinated, maintaining the right conditions is essential for supporting their continued growth. Keep the growing medium consistently moist but not soggy, using a gentle misting or bottom-watering technique to avoid disturbing delicate seedlings. As seedlings develop, gradually introduce them to increased light levels, ensuring they receive adequate exposure without becoming scorched. Thin seedlings as needed to prevent overcrowding, selecting the strongest individuals to continue growing.

Transplanting is the final step in the germination process, transitioning seedlings from their initial environment to their permanent home. Before transplanting, it's important to harden off seedlings by gradually acclimating them to outdoor conditions. This involves exposing them to increasing amounts of sunlight, wind, and temperature fluctuations over a period of one to two weeks. Hardening off reduces transplant shock and prepares seedlings for the challenges of their new environment.

When transplanting, handle seedlings with care to avoid damaging their fragile roots. Use a dibber or your fingers to create a hole in the soil, gently placing the seedling at the same depth it was growing in its original container. Lightly firm the soil around the base to provide stability, and water thoroughly to help settle the soil and eliminate air pockets. With proper care and attention, transplanted seedlings will quickly establish themselves and continue to grow into healthy, vigorous plants.

The art of successful germination is a delicate balance of environmental factors, timing, and technique. By understanding the needs of your seeds and creating an optimal environment for their growth, you can unlock the potential within each tiny seed and watch as it transforms into a flourishing plant. This process not only enriches your garden but also deepens your connection to the natural world, offering a sense of wonder and fulfillment as you nurture life from its very beginnings.

Stratification and Scarification Explained

Stratification and scarification are two horticultural techniques that play a crucial role in breaking seed dormancy and fostering successful germination. For many plant species, especially those adapted to temperate climates, these processes are vital for mimicking natural environmental conditions that seeds would encounter in the wild. By understanding and applying these techniques, gardeners can significantly improve germination rates, bringing a wider variety of plants to life.

In the wild, seeds often enter a period of dormancy, a survival mechanism that ensures they do not germinate under unfavorable conditions. Dormancy can be triggered by various factors, including temperature, moisture, and light. Stratification and scarification address these barriers, essentially tricking seeds into believing that the conditions are right for growth.

Stratification involves exposing seeds to a period of cold, moist conditions, simulating the winter environment many seeds experience before germinating in the spring. This cold treatment is essential for seeds that require a prolonged chill to break dormancy and stimulate the germination process. Stratification can be accomplished through several methods, depending on the resources available and the specific needs of the seeds.

One common method is the refrigeration stratification technique. Seeds are mixed with a damp medium, such as sand, peat moss, or vermiculite, and placed in a plastic bag or container. The mixture is then stored in the refrigerator for a period ranging from a few weeks to several months, depending on the species. This process mimics the natural winter conditions, gradually breaking down the seed's dormancy barriers. It is crucial to monitor moisture levels during this period, ensuring the medium remains slightly damp but not waterlogged.

Outdoor stratification is another approach, where seeds are sown directly into the soil or containers and left to experience the natural winter cycle. This method is particularly effective for native plants that are well-adapted to local climates. However, it requires careful planning to protect seeds from predation and extreme weather conditions. Covering the seeded area with mulch or a protective screen can help shield seeds while allowing them to experience the necessary cold exposure.

Scarification, on the other hand, involves physically altering the seed coat to overcome dormancy. Many seeds possess a hard, impermeable outer shell that prevents water and gases from reaching the embryo. Scarification methods aim to weaken this barrier, facilitating the absorption of moisture and the exchange of gases required for germination.

Mechanical scarification is one technique, where seeds are abraded using tools such as sandpaper, a file, or a knife. By gently rubbing or nicking the seed coat, you can create small openings that allow water to penetrate. It's important to exercise caution and precision during this process to avoid damaging the embryo inside. Mechanical scarification is particularly effective for seeds with thick, tough coats, such as morning glories and sweet peas.

Another scarification method is thermal treatment, which involves exposing seeds to hot water or fluctuating temperatures to break dormancy. For hot water treatment, seeds are soaked in hot (but not boiling) water for a specific duration, allowing the heat to soften the seed coat. This technique is commonly used for seeds like beans and legumes. Temperature fluctuation can also be achieved by alternating between warm and cold conditions, simulating the natural freeze-thaw cycles that many seeds encounter in their native habitats.

Chemical scarification is another approach, using substances like sulfuric acid to erode the seed coat. This method is more complex and should be approached with caution, as it involves handling corrosive chemicals. It's typically reserved for large-scale propagation or for seeds that have proven resistant to other scarification techniques.

Combining stratification and scarification may be necessary for certain seeds that require both cold exposure and physical alteration to break dormancy.

Understanding the specific requirements of each plant species is crucial for determining the most effective approach. Seed packets, cultivation guides, and botanical resources provide valuable insights into the dormancy-breaking needs of various plants.

For beginners, starting with seeds that are known to respond well to stratification and scarification can provide valuable experience and confidence. Over time, experimenting with different techniques and observing the results will deepen your understanding of seed biology and propagation.

Once seeds have undergone stratification or scarification, timely planting is essential to capitalize on the dormancy-breaking effects. Carefully monitor seeds for signs of germination, such as swelling or cracking, and be prepared to sow them promptly. Providing optimal conditions for growth, including appropriate temperature, moisture, and light, will support the transition from seed to seedling.

The techniques of stratification and scarification offer a fascinating glimpse into the adaptability and resilience of plant life. By mastering these methods, gardeners can unlock the potential of a diverse array of plant species, expanding their horticultural repertoire and contributing to the preservation of botanical diversity. As you cultivate your skills, you'll discover the joy and satisfaction of nurturing seeds through their journey from dormancy to vibrant life, enriching both your garden and your connection to the natural world.

Seedling Care and Transplanting

Nurturing seedlings through their early growth stages and preparing them for transplanting into the garden or larger containers is a rewarding phase in the horticultural journey. This delicate period requires careful attention to detail and a gentle hand, as seedlings are particularly vulnerable to environmental stresses and mishandling. With the right care and techniques, you can guide your seedlings towards robust health and ensure they thrive once transplanted.

Once seeds have germinated, the first step in seedling care is to ensure they receive adequate light. Seedlings tend to grow rapidly, and inadequate light can result in leggy, weak plants as they stretch towards the light source. For optimal growth, provide seedlings with 12 to 16 hours of light each day. In many cases, natural sunlight may not be sufficient, especially during the shorter days of winter or early spring. A grow light, placed a few inches above the seedlings, can provide the necessary light spectrum and intensity. Adjust the height of the light as seedlings grow, keeping it close enough to prevent stretching without causing heat damage.

Watering is another critical aspect of seedling care. Seedlings require consistent moisture, but overly wet conditions can lead to damping-off, a fungal disease that causes seedlings to wilt and die. The key is to maintain even moisture by watering from the bottom or gently misting the soil surface, taking care not to oversaturate the soil. Allow the top layer of the growing medium to dry slightly between waterings to reduce the risk of disease. As seedlings develop, they may require more frequent watering to support their increasing size and metabolic needs.

Nutrients play a vital role in supporting the growth and development of seedlings. Most seed-starting mixes contain limited nutrients, so it's important to supplement with a diluted, balanced fertilizer once seedlings develop their first true leaves. These are the leaves that appear after the initial cotyledon leaves and indicate that the seedling is ready to begin taking up nutrients. A water-soluble fertilizer, applied at half strength, can provide the essential nutrients that seedlings need to build strong roots and sturdy stems. Be cautious not to over-fertilize, as this can lead to nutrient burn and stunted growth.

Temperature and airflow are additional factors to consider in seedling care. Seedlings generally prefer temperatures between 65 and 75 degrees Fahrenheit, which can be maintained with the help of a heat mat or by placing seedlings in a warm, draft-free location. Good air circulation is crucial for preventing fungal diseases and encouraging strong stem development. A small fan set on low can gently move air around the seedlings, reducing humidity and strengthening stems by simulating natural breezes.

As seedlings grow, thinning may be necessary to prevent overcrowding and ensure that each plant has adequate space and resources to thrive. Thinning involves removing the weaker seedlings, allowing the strongest to continue growing. This can be done by carefully snipping the unwanted seedlings at the soil level with scissors, minimizing root disturbance to the remaining plants. By giving each seedling sufficient space, you promote healthier growth and reduce competition for light, water, and nutrients.

Hardening off is a crucial step in preparing seedlings for transplanting. This process involves gradually acclimating seedlings to outdoor conditions, reducing the shock of transplanting and helping them adjust to fluctuations in temperature, wind, and sunlight. Begin hardening off about one to two weeks before transplanting by placing seedlings outside in a sheltered location for a few hours each day, gradually increasing their exposure to direct sunlight and outdoor conditions. This gradual transition helps seedlings develop resilience and adapt to their new environment.

When it comes time to transplant seedlings, timing is key. Wait until the risk of frost has passed and soil temperatures are suitable for planting. Choose a cloudy day or transplant in the late afternoon to minimize stress on the seedlings. Before transplanting, water seedlings thoroughly to ensure they are well-hydrated and less prone to transplant shock. Prepare the planting site by loosening the soil and incorporating organic matter, creating a hospitable environment for root growth.

Handle seedlings with care during transplanting to avoid damaging their delicate roots. Gently remove seedlings from their containers, supporting the root ball and minimizing root disturbance. If seedlings are growing in biodegradable pots, you can plant them directly into the soil, but be sure to tear or remove any part of the pot that may impede root growth. Place each seedling into a pre-dug hole at the same depth it was growing in its container, and gently firm the soil around the base to provide stability.

Once transplanted, water seedlings thoroughly to help settle the soil and eliminate air pockets around the roots. A layer of mulch can help retain soil moisture, suppress weeds, and regulate soil temperature, creating an optimal environment for seedling growth. Continue to monitor the seedlings closely, providing additional water and nutrients as needed to support their establishment.

The transition from seedling to mature plant is a testament to the resilience and adaptability of life. By mastering the art of seedling care and transplanting, you play an essential role in this transformation, nurturing young plants through their most vulnerable stages and setting the foundation for a bountiful garden. This journey not only enhances your horticultural skills but also deepens your connection to the rhythms of nature, offering a sense of fulfillment and wonder as you witness the fruits of your labor take root and flourish.

Challenges and Solutions in Seed Propagation

Navigating the world of seed propagation can be as challenging as it is rewarding. While the process of coaxing life from a seemingly inert seed is one of nature's most fascinating feats, it's also fraught with potential pitfalls. Understanding the common challenges faced in seed propagation and the solutions to these hurdles can empower both novice and experienced gardeners to cultivate a thriving garden.

One of the first hurdles in seed propagation is achieving successful germination. Many factors can contribute to low germination rates, including poor seed quality, improper planting depth, and inadequate environmental conditions. To address these issues, start by selecting high-quality seeds from reputable sources. Check for signs of viability such as plumpness and uniform color, and avoid seeds that appear shriveled or discolored. Ensure that seeds are planted at the correct depth, generally two to three times their diameter, unless specific instructions indicate otherwise. Maintaining optimal temperature and moisture levels is also crucial; a seedling heat mat and humidity dome can help create a conducive environment for germination.

Damping-off is another common problem that can sabotage seed propagation efforts. This fungal disease affects seedlings, causing them to wilt and collapse at the soil line. To prevent damping-off, use a sterile, well-draining seed-starting mix and avoid overwatering. Watering from the bottom or using a fine mist can help maintain even moisture without saturating the soil. Good air circulation is also important; consider using a small fan to keep air moving around seedlings and reduce humidity.

Pests pose another challenge, as they can damage or devour tender seedlings before they have a chance to establish themselves. Common pests include aphids, slugs, and fungus gnats. To protect your seedlings, implement an integrated pest management approach. This may involve physical barriers such as row covers, natural predators like ladybugs, or organic solutions such as neem oil or diatomaceous earth. Vigilant monitoring and early intervention can prevent pest infestations from getting out of hand.

Nutrient deficiencies can also hinder seedling growth, manifesting as yellowing leaves, stunted growth, or poor root development. Ensuring that seedlings receive adequate nutrition is essential for their health and vigor. Begin feeding seedlings with a diluted, balanced fertilizer once they develop their first true leaves. If symptoms of deficiency appear, such as chlorosis or purpling of leaves, adjust nutrient levels accordingly. Soil tests can provide valuable insights into nutrient availability and help guide your fertilization strategy.

Environmental stressors, such as temperature fluctuations or inadequate light, can impact seedling development.

Seedlings require consistent temperatures between 65 and 75 degrees Fahrenheit and sufficient light to prevent legginess. If natural sunlight is insufficient, supplement with grow lights to provide the necessary spectrum and intensity. Position the lights close to seedlings, adjusting as they grow to maintain an optimal distance. Protect seedlings from temperature extremes, drafts, and sudden changes by placing them in a stable environment.

Transplant shock is another hurdle that can affect seedlings as they transition from indoor conditions to the garden. Hardening off is a critical step in minimizing transplant shock. Gradually acclimate seedlings to outdoor conditions by increasing their exposure to sunlight, wind, and temperature fluctuations over a period of one to two weeks. When transplanting, choose a cool, overcast day or transplant in the late afternoon to reduce stress. Ensure that seedlings are well-watered before and after transplanting, and consider using a root stimulator to encourage strong root growth in their new location.

Timing is a crucial aspect of seed propagation that can influence success. Sowing seeds too early or late can result in poor germination or growth. Consider the specific requirements of each plant species, including the length of the growing season and the local climate. Utilize planting calendars and guides to determine the best time to sow seeds, and be mindful of factors such as frost dates and temperature patterns.

Space constraints can also pose a challenge, particularly for those gardening in urban or limited environments. Innovative solutions such as vertical gardening, container gardening, or utilizing raised beds can maximize available space and provide ideal conditions for seed propagation. Selecting compact or dwarf plant varieties can further enhance space efficiency, allowing you to cultivate a diverse array of plants even in small areas.

Record-keeping is an often-overlooked aspect of seed propagation that can greatly aid in troubleshooting challenges and refining techniques. Maintaining a gardening journal with details about seed varieties, planting dates, germination rates, and any issues encountered can provide valuable insights over time. This record can serve as a reference for future plantings, helping you adjust strategies and improve outcomes.

By understanding the challenges and solutions in seed propagation, gardeners can approach this rewarding endeavor with confidence and resilience. Each obstacle presents an opportunity for learning and growth, enriching your horticultural journey and deepening your connection to the natural world. With perseverance and adaptability, you can transform potential setbacks into stepping stones, cultivating not only a flourishing garden but also a greater appreciation for the intricate dance of life that unfolds with each seed sown.

Chapter 3: Vegetative Propagation Techniques

The Art of Cutting: Stems, Leaves, and Roots

The art of propagation through cuttings—whether stems, leaves, or roots—offers gardeners a fascinating and efficient method of multiplying plants. This technique not only allows for the creation of exact genetic replicas, or clones, of a parent plant but also provides an opportunity to cultivate new plants without the need for seeds. Success in this endeavor requires a blend of knowledge, precision, and a touch of creativity.

The journey begins by selecting the appropriate plant and cutting type. Certain plants lend themselves more readily to propagation through cuttings, with popular candidates including houseplants, shrubs, and perennials. Understanding the type of cutting—stem, leaf, or root—that will work best with the plant species is essential. For instance, woody plants often propagate well from stem cuttings, while succulents may thrive from leaf cuttings.

Stem cuttings are perhaps the most commonly used method and are typically taken from healthy, non-flowering shoots. The best time for collecting stem cuttings is during the plant's active growth phase, when stems are firm yet still flexible. Using a sharp knife or pruning shears, cut a section of stem about four to six inches long, just below a node where leaves attach. This node is crucial, as it houses the cells that will develop into roots.

Preparation of the stem cutting involves removing the lower leaves and any flowers or buds, leaving a few leaves at the top. This reduces water loss through transpiration and directs the plant's energy towards root development. If the plant is particularly leafy, consider cutting the remaining leaves in half to minimize moisture loss further.

Rooting hormones can enhance the success rate of stem cuttings, particularly for plants that are more challenging to propagate. These hormones, available in powder, liquid, or gel form, stimulate root cell development. Dip the cut end of the stem into water, then into the rooting hormone, tapping off any excess. Although some plants root successfully without these aids, using them can offer an edge, especially for beginners.

The choice of growing medium is pivotal. A well-draining mix, such as a combination of sand, perlite, and peat moss, provides the right balance of moisture and aeration. Fill a pot with the medium, and use a pencil or stick to make a hole for the cutting, ensuring the rooting hormone remains undisturbed. Insert the cutting into the hole and firm the medium around it to provide stability.

Humidity and temperature play a vital role in root development. Stem cuttings thrive in warm, humid conditions, which can be simulated by placing a plastic bag or a propagator lid over the pot. This creates a mini-greenhouse effect, retaining moisture and warmth. Position the pot in a bright location,avoiding direct sunlight that could overheat or scorch the cuttings. Maintaining a consistent temperature of around 70 to 75 degrees Fahrenheit is ideal for root development.

Leaf cuttings offer another avenue for propagation, particularly suited to certain houseplants and succulents. This method requires selecting a healthy, mature leaf and carefully detaching it from the plant with as little damage as possible. For succulents, ensure the leaf separates cleanly at the base, as any remaining stem tissue can hinder rooting. Allow the leaf to dry for a day or two, forming a callus over the cut edge to prevent rot.

Once the leaf has callused, place it on a well-draining medium, with the cut edge slightly buried. For some plants, such as African violets or begonias, the leaf may be cut into sections, each capable of producing new plants. Similar to stem cuttings, maintaining humidity is crucial. A plastic cover can help preserve moisture and warmth, encouraging new growth.

Root cuttings are less common but offer unique advantages, particularly for plants with spreading root systems, such as raspberries and horseradish. This method involves carefully digging up a section of the plant's roots during dormancy, typically in late fall or early spring. Select healthy roots, and cut them into sections two to four inches long. Plant these horizontally in a suitable medium, and cover them lightly with soil. Keep the cuttings moist and in a cool location until new shoots emerge.

An often overlooked aspect of successful propagation is patience. Rooting can take anywhere from a few weeks to several months, depending on the plant species and environmental conditions. Regularly check the cuttings for signs of root development by gently tugging on them; resistance indicates that roots have formed. Once roots are established, gradually acclimate the young plants to less humid conditions by removing the cover for increasing periods each day.

Transplanting rooted cuttings requires care to avoid damage to the delicate new roots. Gently remove the cutting from the medium, preserving as much of the root ball as possible. Pot the cutting into a suitable growing medium, and water thoroughly to help the soil settle around the roots. As the plant matures, continue to provide optimal conditions of light, water, and nutrients to encourage robust growth.

Through the art of cutting propagation, gardeners can cultivate a diverse array of plants, from beloved houseplants to cherished garden perennials. This method not only expands one's gardening repertoire but also fosters a deeper connection with the plants themselves, as each cutting represents a new chapter in the life of the plant.

Embrace the process with curiosity and creativity, and watch as a single snip brings forth a world of growth and possibility.

Layering Methods for Plant Propagation

Layering is a fascinating and effective plant propagation technique that has been used for centuries to produce new plants from existing ones. Unlike other propagation methods that require cutting or detaching parts of the plant, layering involves encouraging roots to form on a stem while it is still attached to the parent plant. This approach ensures that the developing plant continues to receive nourishment and support until it is robust enough to survive on its own. Layering is particularly advantageous for propagating woody shrubs, vines, and certain perennials, offering a reliable way to expand your garden with minimal risk.

One of the most straightforward and widely practiced methods is simple layering. This technique is especially effective for flexible, low-growing branches. To begin, select a healthy, vigorous branch on the parent plant. Bend the branch down to the ground, and identify a point where it can be buried in the soil while still attached to the main plant. At this point, make a slight wound or nick on the underside of the branch using a sharp knife. This injury encourages root formation by exposing the cambium layer, the part of the plant responsible for producing new cells.

Once the wound is made, dig a shallow trench and secure the wounded section of the branch into it, using soil to cover the area. A U-shaped pin or a small rock can help hold the branch in place. Ensure the tip of the branch remains exposed to continue growing upward. Over time, roots will develop from the wounded area, anchoring the new plant into the soil. This process can take several weeks to a few months, depending on the plant species and environmental conditions.

Air layering offers an intriguing alternative for plants with rigid or upright stems that cannot be easily bent to the ground. This method is often used for indoor plants, shrubs, and trees. To execute air layering, select a healthy stem on the parent plant and make a shallow cut or remove a ring of bark around the stem at the desired rooting point. This girdling interrupts the flow of nutrients, stimulating the formation of roots above the cut.

After preparing the stem, apply a rooting hormone to the exposed area to promote root growth. Encase the wounded section in a moist medium, such as sphagnum moss, and wrap it with clear plastic to retain moisture. Secure the plastic with twist ties or string at both ends, creating a sealed environment around the wound. Monitor the moisture level within the plastic wrap, adding water as needed to keep the moss damp. In time, roots will emerge within the moss, visible through the clear plastic.

Once a healthy root system has developed, the layered stem can be removed from the parent plant. Carefully cut below the root mass, and transplant the new plant into a suitable potting mix or directly into the garden. Provide gentle care and acclimatization to ensure successful establishment in its new location.

Mound layering, also known as stooling, is particularly effective for propagating shrubs and trees that produce multiple shoots from a central base. This method involves cutting back the parent plant to encourage the growth of new shoots. As these shoots emerge and grow, mounding soil around their base encourages root formation at the buried sections.

Begin mound layering in early spring by cutting the parent plant back to ground level. As new shoots develop, gradually build up a mound of soil around their base, burying a portion of each shoot. Over the growing season, roots will form at the soil-covered nodes. By autumn or the following spring, these rooted shoots can be separated from the parent plant and transplanted as independent specimens.

Trench layering is another method well-suited for plants with long, flexible branches. This technique is similar to simple layering but involves burying the entire length of a branch horizontally in a trench. As the branch is buried, upward-facing buds will grow into new shoots, each capable of forming its own root system.

To perform trench layering, dig a trench near the base of the parent plant, and lay a selected branch along its length. Cover the branch with soil, ensuring that the tips of the buds remain exposed above the soil surface. Secure the branch in place, and water the trench to encourage root and shoot development. Over time, each bud will sprout into a new plant, which can be separated from the parent by cutting along the buried branch.

Despite the simplicity and success of layering methods, patience is essential, as the process of root development can be slow. However, the advantages of layering, such as a higher success rate and less stress on the plant, make it a valuable technique in the gardener's toolkit. By mastering these methods, you can propagate a wide range of plants, expanding your garden with genetically identical specimens and preserving the characteristics of cherished plants.

Layering not only enriches your garden but also deepens your understanding of plant growth and reproduction. It connects you to the natural processes that allow plants to thrive and multiply, offering a sense of fulfillment as you witness the transformation from a single branch to a thriving new plant. Embrace the art and science of layering, and enjoy the rewards of this ancient and enduring propagation technique.

Division of Perennials and Bulbs

Dividing perennials and bulbs stands as a time-honored practice in gardening, offering both rejuvenation for plants and multiplication for gardens. As these plants mature and clump together, they often become crowded, leading to reduced vigor and fewer blooms. Division helps alleviate this congestion, encouraging healthy growth and flowering while providing gardeners with additional plants to expand their landscapes or share with friends.

Understanding the timing and method for division is key to success. Typically, perennials are divided every three to five years, depending on their growth rate and specific needs. The best time to divide depends on the plant's blooming period. Spring and summer-flowering perennials are usually divided in the fall, while fall-flowering plants are best divided in the spring. This timing ensures that the plant can recover and establish itself before the next blooming season.

Before beginning the division process, prepare the new planting sites by amending the soil with organic matter such as compost or well-rotted manure. This enriches the soil, providing a nutrient-rich environment that supports the newly divided plants. Water the plants thoroughly a day or two before division, as this helps reduce stress and makes digging easier.

To divide a perennial, start by cutting back the foliage to about six inches. This reduces water loss and makes handling the plant more manageable. Using a spade or garden fork, carefully dig around the plant's root ball, maintaining a generous distance from the crown to avoid damaging the roots. Gently lift the plant from the ground, shaking off excess soil to expose the root system.

Inspect the roots for signs of overcrowding, rot, or pests. Healthy roots are firm and white, while damaged or diseased sections should be discarded. Use a sharp knife, spade, or your hands to separate the plant into smaller sections, ensuring each division has a portion of the crown and an adequate root system. If the root mass is particularly tough, two garden forks can be inserted back-to-back to pry the plant apart.

Once divided, it's time to replant the divisions in their new locations. Space them according to the mature size of the plant, allowing for ample growth without future overcrowding. Plant each division at the same depth it was growing previously, and water thoroughly to help settle the soil and eliminate air pockets. Applying a layer of mulch can help retain moisture and suppress weeds, fostering a healthy start for the new plants.

Bulbs, much like perennials, benefit from periodic division to maintain their vigor and proliferation. Over time, bulbs can become crowded, leading to smaller blooms and less robust plants. Division not only remedies this but also increases your stock of bulbs, allowing the beauty of your garden to spread.

The timing for dividing bulbs depends on the type of bulb and its flowering period. Spring-flowering bulbs, such as daffodils and tulips, are best divided in late summer or early fall. Summer-flowering bulbs, like lilies and gladiolus, are divided in the fall or early spring. Wait until the foliage has died back naturally, as this allows the bulb to store energy for the next growing season.

To divide bulbs, carefully dig up the clump, taking care not to damage the bulbs. Gently shake off the soil to reveal the individual bulbs. Inspect each one, discarding any that are soft, discolored, or damaged. Separate the bulbs by hand, ensuring each has a healthy basal plate and is free from disease.

Replant the bulbs in well-draining soil, spacing them according to their specific requirements. Plant them at a depth approximately two to three times their height, with the pointed end facing upwards. Water them well after planting, and apply mulch to protect the bulbs and retain soil moisture.

Division is not only a practical horticultural technique but also an act of renewal and growth. It allows gardeners to refresh their gardens, ensuring that perennials and bulbs continue to thrive and bring beauty year after year. Moreover, it offers the joy of sharing, as divisions can be given to fellow gardeners, spreading the love of plants and gardening far and wide.

As you engage in the division of your perennials and bulbs, you are participating in a cycle of life that has sustained gardens for generations. Embrace this opportunity to nurture your plants, expand your garden, and share your passion with others. With each division, you contribute to the legacy of your garden, enriching it with vitality and promise for the future.

Grafting and Budding Techniques

Grafting and budding stand as sophisticated yet accessible horticultural techniques that have been used by gardeners and orchardists for centuries to improve plant vigor, yield, and resilience. These methods involve joining two plants together in such a way that they grow as one, allowing the desirable traits of both plants to be combined. By mastering these techniques, gardeners can not only propagate plants with superior characteristics but also preserve rare and heirloom varieties.

Grafting involves the union of two plant segments: the rootstock, which forms the base of the plant, and the scion, which is the desired plant variety grafted onto the rootstock. This method is particularly beneficial for fruit trees and woody ornamentals, where it can enhance disease resistance, increase cold hardiness, and improve overall growth performance.

Selecting the right rootstock is a crucial first step in the grafting process. Rootstocks are chosen for their adaptability to soil conditions, disease resistance, and growth habits. They provide the foundation for the grafted plant, influencing its size, vigor, and longevity. Once the appropriate rootstock is selected, it's time to choose a scion that possesses the desired fruiting or flowering characteristics.

The timing of grafting plays a significant role in its success. Most grafting is done in late winter or early spring, when both the rootstock and scion are dormant but just before the sap begins to flow. This timing allows the graft to heal and establish a strong union as the plant awakens and begins its growth cycle.

Several grafting techniques exist, each suited to specific plants and conditions. One of the most common methods is the whip-and-tongue graft, which provides a strong union and is often used for fruit trees. To perform this graft, both the rootstock and scion are cut at a matching angle, creating a flat surface for joining. A tongue is then cut into each piece, allowing them to interlock securely. Once aligned, the graft is wrapped tightly with grafting tape or a rubber band to hold the pieces together and protect the union from drying out.

Cleft grafting, another widely used method, is ideal for larger rootstocks or when the scion is significantly smaller than the rootstock. To execute a cleft graft, the rootstock is cut horizontally and then split vertically to create a cleft. The scion, prepared with a sharply tapered end, is inserted into the cleft, ensuring that the cambium layers of both pieces align closely. The graft is then sealed with grafting wax or a similar compound to prevent moisture loss and infection.

Side grafting, often used for conifers and certain ornamentals, involves making a slanted cut on the side of the rootstock and inserting a similarly cut scion. This method allows for the graft to be positioned on the side, facilitating healing and growth.

Budding, a variation of grafting, involves the use of a single bud from the desired plant instead of a larger scion. This technique is particularly useful for propagating roses and fruit trees, offering a high success rate with minimal scion material. T-budding is one of the most popular budding methods, performed during the growing season when the bark of the rootstock is slipping, meaning it can be easily lifted.

To perform T-budding, a T-shaped cut is made in the bark of the rootstock, and the bud is inserted into the flap created by the cut. The bud is then secured with budding tape, ensuring it remains in place as it heals and begins to grow. Chip budding, another effective method, involves cutting a chip of bark and wood from both the rootstock and scion and matching them precisely before securing them with tape.

The success of both grafting and budding relies heavily on proper aftercare. The grafted plant must be protected from environmental stresses, such as wind, excessive sun, and pests. Maintaining moisture levels is critical; the area around the graft should remain hydrated but not waterlogged. As the graft heals and begins to grow, any shoots or growth below the graft union should be removed to prevent competition with the scion or bud.

Monitoring the graft for signs of success or failure is an ongoing process. A successful graft will show signs of growth, such as swelling buds or new shoots, within a few weeks. If the graft fails, it may be due to misalignment, poor union, or environmental factors, and the process may need to be repeated with adjustments to technique or timing.

Grafting and budding are not only practical techniques for plant propagation but also artistic endeavors that connect gardeners with the rich traditions of horticulture. Through these methods, plants are shaped and improved, yielding fruit, flowers, and beauty that might otherwise be unattainable. As you practice and refine your grafting and budding skills, you contribute to the legacy of your garden, blending science and artistry in the pursuit of botanical excellence.

Troubleshooting Common Vegetative Propagation Issues

Vegetative propagation offers gardeners a wonderful and efficient way to produce new plants, but like any gardening technique, it comes with its own set of challenges. Understanding and addressing these common issues can make the difference between thriving plants and frustrating failures. By learning to troubleshoot effectively, you can ensure a successful propagation experience and a flourishing garden.

A frequent issue in vegetative propagation is poor rooting, which can manifest as slow or nonexistent root development. This problem often stems from inadequate environmental conditions. Temperature plays a crucial role; cuttings typically require warmth to encourage root formation. If your propagation area is too cold, consider using a heat mat to maintain a consistent temperature around 70 to 75 degrees Fahrenheit. Additionally, ensure that your growing medium remains consistently moist—neither too dry nor waterlogged. Overwatering can lead to root rot, while underwatering can cause desiccation and failure to root.

The choice of rooting medium is another factor that can impact success. A well-draining medium, such as a mix of perlite, sand, and peat moss, provides the right combination of moisture retention and aeration. If your medium is too dense or retains excess water, consider amending it to improve drainage. Also, ensure that your cutting tools are sharp and clean, as dull or dirty tools can crush plant tissues, leading to increased susceptibility to disease.

Damping-off, a disease caused by soil-borne fungi, is a notorious problem for young seedlings and cuttings. It often appears as a sudden wilting and collapse of the plant at the soil line. To prevent damping-off, always use sterile containers and a fresh, sterile growing medium. Good air circulation is essential; consider placing a fan near your propagation area to enhance ventilation. Water your cuttings from the bottom when possible, and avoid overhead watering, which can create a humid environment conducive to fungal growth.

Pests can also be a significant hindrance to successful propagation. Aphids, spider mites, and fungus gnats are common culprits that can damage tender cuttings. Regularly inspect your plants for signs of infestation, such as webbing, distorted leaves, or sticky residue. Introduce beneficial insects like ladybugs or predatory mites to control pest populations naturally. Additionally, neem oil or insecticidal soap can be effective treatments for many common pests without harming the plants.

Nutrient deficiencies may manifest as yellowing leaves, stunted growth, or poor root development. During the initial stages of propagation, cuttings rely on stored nutrients, but as they develop roots, they require external nutrition. Once roots are established, begin feeding your plants with a diluted, balanced fertilizer to provide the necessary nutrients for healthy growth. If you notice specific deficiency symptoms, such as chlorosis or purpling, adjust your fertilization regimen accordingly.

Light conditions play a crucial role in the success of vegetative propagation. Insufficient light can lead to leggy, weak plants, while too much direct sunlight can scorch tender cuttings. Position your cuttings in a location with bright, indirect light. If natural light is inadequate, supplement with grow lights to ensure your plants receive the necessary light spectrum and intensity. Adjust the distance between the lights and the plants to prevent overheating and ensure even growth.

Transplant shock is another hurdle to overcome when moving propagated plants from their propagation environment to their permanent location. Minimize shock by gradually acclimating your plants to outdoor conditions, a process known as hardening off. Start by placing the plants in a sheltered outdoor area for a few hours each day, gradually increasing their exposure to sunlight, wind, and temperature fluctuations over a week or two. Water the plants well before and after transplanting to reduce stress and help them establish in their new environment.

Timing is a critical aspect of successful propagation that can often be overlooked. Initiating propagation at the right time of year, when the parent plant is in its active growth phase, can significantly improve your chances of success.

Consult planting guides and calendars to determine the best time to propagate specific plant species, and consider factors such as local climate and growing season length.

Record-keeping is an invaluable tool for troubleshooting and refining your propagation techniques. Maintain a detailed journal of your propagation efforts, including information about the plant species, cuttings taken, environmental conditions, and any issues encountered. This record can provide valuable insights into what works and what doesn't, allowing you to make informed adjustments to your methods over time.

By understanding the common challenges of vegetative propagation and implementing practical solutions, you can transform potential setbacks into opportunities for learning and growth. Each issue you encounter provides valuable experience, deepening your knowledge and honing your skills as a gardener. Embrace the process with curiosity and resilience, and watch as your garden flourishes with the fruits of your labor.

Chapter 4: Advanced Propagation Strategies

Micropropagation and Tissue Culture

Micropropagation and tissue culture represent the pinnacle of modern plant propagation techniques, offering unparalleled precision and efficiency in producing large numbers of genetically identical plants. These methods utilize small plant tissues or cells under sterile conditions to cultivate new plants, making them invaluable for both research and commercial applications. For beginners, understanding the basics of these techniques can open new avenues in gardening, agriculture, and horticulture.

The journey into micropropagation begins with the selection of the explant, a small piece of plant tissue used to initiate the culture. Explants can be taken from various parts of the plant, such as leaves, stems, buds, or even roots, depending on the species and the desired outcome. The choice of explant is crucial, as it influences the success rate and quality of the resulting plants. It's essential to select healthy, disease-free plant material to minimize the risk of contamination.

Once the explant is chosen, it must be sterilized to eliminate any microorganisms that could compromise the culture. This process typically involves soaking the explant in a series of sterilizing solutions, such as bleach or ethanol, followed by thorough rinsing with sterile water. Maintaining a sterile environment is paramount throughout the micropropagation process, as even a single contaminant can rapidly overtake the culture.

The sterilized explant is then placed on a nutrient-rich medium, which provides the essential minerals, vitamins, and hormones required for growth. This medium is usually gel-based, allowing the explant to remain stationary while absorbing nutrients. The composition of the medium can be tailored to promote specific developmental stages, such as callus formation, shoot induction, or root initiation.

Micropropagation progresses through several distinct stages. Initially, the explant may develop a callus, a mass of undifferentiated cells that serves as a precursor to organ formation. With the application of specific plant hormones, such as cytokinins and auxins, the callus can be induced to form shoots or roots. Cytokinins generally promote shoot development, while auxins encourage root formation. The ability to manipulate these hormones allows for precise control over the plant's development, making micropropagation a versatile tool for plant cultivation.

As the culture develops, the newly formed shoots are transferred to fresh media to encourage further growth and root formation. This stage, known as the multiplication phase, is where the power of micropropagation truly shines, as each shoot can be subdivided and cultured independently, resulting in exponential plant production. This capability is particularly beneficial for mass-producing rare or endangered plant species, as well as high-value crops and ornamentals.

Rooted plantlets are eventually acclimatized to non-sterile conditions, a critical step in transitioning them from the controlled laboratory environment to natural settings. This process, known as hardening off, involves gradually exposing the plantlets to ambient humidity, light, and temperature conditions to build their resilience. Typically, plantlets are first transferred to a greenhouse or controlled nursery setting, where conditions can be carefully monitored and adjusted.

Micropropagation offers several advantages over traditional propagation methods. The ability to produce large numbers of uniform plants in a relatively short time frame is a significant benefit, especially for species that do not produce viable seeds or have long dormancy periods. Additionally, because micropropagation is conducted in a sterile environment, it minimizes the risk of disease transmission, resulting in healthier plants.

Despite its advantages, micropropagation does present certain challenges. The need for sterile conditions requires specialized equipment and facilities, which can be a barrier for small-scale growers. Moreover, maintaining the precise balance of nutrients and hormones in the culture medium demands expertise and attention to detail. Nonetheless, with proper training and resources, these challenges can be overcome, making micropropagation a viable option for many growers.

Tissue culture, a broader term encompassing micropropagation, also includes techniques such as embryo culture and protoplast culture. Embryo culture involves the cultivation of plant embryos in vitro, often used to rescue embryos from hybrid crosses that might otherwise fail to develop. Protoplast culture, on the other hand, involves the isolation and culture of plant cells without cell walls, enabling the study of cellular processes and genetic manipulation.

The applications of micropropagation and tissue culture extend far beyond simple plant production. These techniques have revolutionized plant breeding, allowing for the rapid development and testing of new cultivars with desirable traits such as disease resistance, improved yield, or enhanced nutritional content. In conservation biology, micropropagation serves as a critical tool for preserving genetic diversity and restoring endangered plant populations.

For beginners interested in exploring micropropagation, starting with a few basic materials and a simple home setup is feasible. Sterile containers, a basic nutrient medium, and a few explants from a familiar plant species, such as African violets or orchids, provide an accessible introduction to the technique. Online resources and workshops can offer guidance on setting up a home lab and refining your skills.

As you gain experience, micropropagation opens doors to a deeper understanding of plant biology and the intricate processes that govern growth and development. It invites you to experiment, innovate, and push the boundaries of traditional gardening practices. Whether you're cultivating rare species, developing new hybrids, or simply expanding your garden, micropropagation and tissue culture offer a world of possibilities limited only by your imagination and curiosity.

Hydroponic and Aquaponic Systems

Hydroponic and aquaponic systems have revolutionized modern agriculture, offering innovative solutions for growing plants without soil. These systems harness the power of water and nutrients to cultivate plants efficiently, making them ideal for both urban agriculture and areas with poor soil quality. For those new to these methods, understanding the fundamentals can open up a world of possibilities for sustainable and productive gardening.

Hydroponics, derived from the Greek words for "water" and "labor," involves growing plants in a nutrient-rich water solution. This method provides precise control over the growing environment, allowing plants to access the exact nutrients they need for optimal growth. The absence of soil reduces the risk of soil-borne diseases and pests, resulting in healthier plants and higher yields.

To set up a hydroponic system, you'll need several key components: a water reservoir, a nutrient solution, a pump to circulate the water, and a growing medium to support the plants. The growing medium, which can be anything from rock wool to clay pellets, provides stability for the plant roots while allowing them to absorb water and nutrients. Depending on the type of hydroponic system you choose, you may also need additional equipment such as grow lights or timers.

There are various types of hydroponic systems, each with its unique advantages. The nutrient film technique (NFT) involves a continuous flow of nutrient solution over the plant roots, providing a consistent supply of nutrients and oxygen. This method is particularly effective for leafy greens and herbs, which thrive in the constant nutrient flow. Another popular system is the deep water culture (DWC), where plant roots are suspended in a nutrient solution, and air is bubbled through the water to provide oxygen. DWC is known for its simplicity and effectiveness, making it a great choice for beginners.

Ebb and flow systems, also known as flood and drain systems, periodically flood the plant roots with nutrient solution and then allow it to drain away. This cycle provides plants with both nutrients and oxygen, promoting healthy growth. This system is versatile and can be adapted for a wide range of plants, from vegetables to flowering ornamentals. Drip systems, on the other hand, deliver a slow, steady supply of nutrient solution directly to the plant roots, making them ideal for larger plants or those with specific nutrient requirements.

Aquaponics combines the principles of hydroponics with aquaculture, the practice of raising fish or other aquatic animals. In an aquaponic system, fish waste provides a natural source of nutrients for the plants, creating a symbiotic relationship between the two. The plants, in turn, filter the water, maintaining a healthy environment for the fish. This closed-loop system is highly sustainable, as it minimizes water usage and waste.

Setting up an aquaponic system involves several components similar to hydroponics, with the addition of a fish tank and a biofilter to convert fish waste into plant nutrients. The choice of fish species is crucial, as different fish have varying environmental and dietary needs. Tilapia, for example, are hardy and adaptable, making them a popular choice for aquaponics. Other species, such as trout or catfish, may require specific temperature or water quality conditions.

Aquaponics offers the advantage of producing both plants and fish, making it a highly efficient system for food production. It can be particularly beneficial in areas with limited resources or access to fresh produce, as it can provide a continuous supply of fresh vegetables and protein. Additionally, the integration of fish and plants creates a balanced ecosystem, reducing the need for synthetic fertilizers and pesticides.

Both hydroponic and aquaponic systems require careful monitoring and maintenance to ensure optimal plant growth and system health. Regularly check the pH and nutrient levels of the water, as imbalances can affect plant health and productivity. In aquaponics, monitoring water quality parameters such as ammonia, nitrite, and nitrate levels is essential to maintain a healthy environment for the fish.

Lighting is another critical factor in these systems, especially if you're growing indoors or in areas with limited natural sunlight. Full-spectrum LED grow lights can provide the necessary light for photosynthesis, allowing plants to grow year-round regardless of external conditions. Adjusting the light intensity and duration based on the plant species and growth stage can optimize growth and yield.

Venturing into hydroponics or aquaponics offers an exciting opportunity to explore cutting-edge agricultural techniques. These systems are not only efficient and sustainable but also adaptable to a wide range of environments and scales. Whether you're looking to grow fresh produce in a small urban apartment or establish a commercial farm, hydroponic and aquaponic systems provide the tools to do so effectively.

The journey into these innovative systems is one of discovery and experimentation. As you gain experience, you'll develop a deeper understanding of plant biology, water chemistry, and ecosystem dynamics. This knowledge empowers you to tailor your system to meet your specific goals, whether they're centered around maximizing yield, reducing environmental impact, or simply enjoying the process of growing your own food. Embrace the challenges and rewards of hydroponics and aquaponics, and watch as your efforts bear fruit—or vegetables—right before your eyes.

Propagation in Controlled Environments

Cultivating plants in controlled environments offers a unique opportunity to manipulate and optimize growing conditions, ensuring successful propagation across diverse species. These environments, ranging from greenhouses to growth chambers, provide the flexibility to regulate factors such as temperature, humidity, light, and airflow, creating ideal conditions for plant growth and development. For those new to propagation in controlled settings, understanding how to harness these variables is crucial for maximizing success and efficiency.

Controlled environments are particularly beneficial for propagating plants that require specific conditions not easily achieved outdoors. By regulating temperature, you can ensure consistent warmth, which is essential for seed germination and rooting of cuttings. Many seeds and cuttings thrive at temperatures between 65 and 75 degrees Fahrenheit, a range that can be easily maintained in a controlled setting. Heating mats or under-bench heating systems can provide additional warmth to the root zone, promoting faster and more uniform root development.

Humidity control is another critical aspect of propagation in controlled environments. High humidity levels help prevent moisture loss from cuttings and young seedlings, reducing the risk of desiccation. However, excessive humidity can promote fungal diseases, so it's important to strike the right balance. Using misting systems or humidifiers can maintain optimal humidity levels, while proper ventilation helps prevent the buildup of excess moisture. Regularly monitoring and adjusting humidity levels ensures a healthy environment for plant growth.

Light is a fundamental factor in plant propagation, influencing photosynthesis and growth. In controlled environments, artificial lighting can supplement or replace natural sunlight, providing plants with the necessary light spectrum and intensity. LED grow lights are particularly effective, as they can be tailored to emit specific wavelengths that promote growth and flowering. Adjusting light duration and intensity based on plant species and growth stage can optimize photosynthesis, leading to vigorous and healthy plants.

Airflow and ventilation are often overlooked but are vital components of a controlled environment. Good air circulation prevents the buildup of heat and humidity, reducing the risk of disease. Fans can be used to provide consistent airflow, ensuring that the growing environment remains stable and conducive to plant health. Additionally, proper ventilation introduces fresh air, supplying the carbon dioxide necessary for photosynthesis.

The choice of growing medium is another critical consideration in controlled environments. Soilless mixes, such as peat-based or coir-based substrates, are commonly used due to their excellent drainage and aeration properties. These mixes can be tailored with additional amendments, such as perlite or vermiculite, to meet specific plant needs. The growing medium should provide stable support for the plants while allowing for efficient nutrient and water uptake.

Nutrient management is an important aspect of propagation in controlled environments. Plants require a balanced supply of essential nutrients for optimal growth, and this can be achieved through the use of liquid or granular fertilizers. In hydroponic systems, nutrient solutions are directly supplied to the plant roots, allowing for precise control over nutrient availability. Regularly testing nutrient levels and adjusting the fertilizer regimen as needed will prevent deficiencies or toxicities, ensuring healthy plant development.

Pest and disease management is crucial in controlled environments, where conditions can inadvertently favor the proliferation of pests and pathogens. Implementing integrated pest management (IPM) strategies helps maintain a healthy growing environment. This approach combines cultural, physical, biological, and chemical methods to manage pest populations. Regularly inspecting plants for signs of infestation or disease, maintaining cleanliness, and introducing beneficial insects are all effective IPM practices.

Propagation in controlled environments offers several advantages over traditional methods. The ability to manipulate environmental conditions allows for year-round cultivation, regardless of external weather patterns. This is especially beneficial for commercial operations, where consistent production is essential. Controlled environments also provide the opportunity to experiment with new plant varieties or techniques, as variables can be precisely adjusted and monitored.

For beginners, starting with a small-scale setup, such as a mini greenhouse or a grow tent, can provide valuable experience with controlled environment propagation. These setups are relatively affordable and easy to manage, offering an accessible introduction to the principles and practices involved. As you gain confidence and expertise, you can expand your setup to accommodate a wider range of plants and techniques.

The flexibility and control offered by these environments empower growers to push the boundaries of traditional gardening, exploring new possibilities and achieving remarkable results. Whether you're cultivating exotic species, developing new hybrids, or simply enjoying the satisfaction of growing your own plants, propagation in controlled environments provides the tools and techniques to succeed. Embrace the challenges and rewards of this innovative approach, and watch as your garden thrives in the perfect balance of science and nature.

The Role of Biotechnology in Modern Propagation

Biotechnology has emerged as a transformative force in modern plant propagation, blending the art of traditional horticulture with cutting-edge scientific advancements. By leveraging biotechnological techniques, we can enhance plant growth, increase yields, and develop new plant varieties with desirable traits. This intersection of technology and nature offers exciting opportunities and challenges, shaping the future of agriculture and plant science.

One of the most significant contributions of biotechnology to propagation is the development of genetically modified organisms (GMOs). Through genetic engineering, specific genes can be introduced, modified, or silenced to confer beneficial traits such as pest resistance, drought tolerance, or improved nutritional content. For example, Bt crops have been engineered to express a natural insecticide derived from Bacillus thuringiensis, providing built-in protection against certain pests and reducing the need for chemical pesticides. This approach not only enhances crop productivity but also supports sustainable agricultural practices.

Another biotechnological advancement in propagation is marker-assisted selection (MAS). This technique involves identifying and selecting plants with desired traits based on genetic markers linked to those traits. MAS accelerates the breeding process, allowing horticulturists to develop new plant varieties more efficiently than traditional methods. By focusing on genetic markers, breeders can ensure that plants possess the desired characteristics, such as disease resistance or improved flavor, without the lengthy trial-and-error process of conventional breeding.

Tissue culture and micropropagation, already explored in previous chapters, also benefit from biotechnological innovations. Advances in culture media formulations, plant hormones, and growth regulators have improved the efficiency and success rates of these techniques. Additionally, biotechnological tools enable the rapid multiplication of plants with minimal starting material, preserving genetic diversity and protecting rare or endangered species.

Biotechnology has also paved the way for somatic embryogenesis, a process in which somatic cells (non-reproductive cells) develop into embryos capable of growing into full plants. This technique allows for large-scale propagation of plants that are difficult to reproduce through conventional means, such as certain forest trees and ornamental species. Somatic embryogenesis is particularly valuable for producing uniform, high-quality plants for commercial purposes, ensuring consistency and reliability in the final product.

The role of biotechnology extends beyond propagation to include the development of novel plant-based products. Through metabolic engineering, scientists can modify plants to produce valuable compounds such as pharmaceuticals, biofuels, and bioplastics. This approach not only diversifies the applications of plants but also contributes to sustainable development by providing renewable alternatives to fossil-based resources.

Despite its many advantages, the integration of biotechnology in plant propagation raises important ethical and environmental considerations. The potential impact of GMOs on ecosystems, biodiversity, and human health remains a topic of debate and research. Regulatory frameworks and risk assessments play a critical role in ensuring that biotechnological advancements are safe, sustainable, and socially acceptable.

For those interested in exploring biotechnology in propagation, education and awareness are key. Understanding the science behind these techniques, as well as the broader implications, empowers individuals to make informed decisions about their use and application. Engaging with reputable sources, participating in workshops or courses, and staying updated on the latest research can provide valuable insights into the world of biotechnological propagation.

The role of biotechnology in modern propagation underscores the dynamic nature of plant science, where tradition and innovation intersect to create new possibilities. As we navigate the complexities and opportunities of this field, we are reminded of the incredible potential of plants and the importance of responsible stewardship. Embracing biotechnology in propagation is not just about enhancing productivity or aesthetics; it's about harnessing the power of nature to address global challenges and improve the quality of life for future generations.

Innovations and Future Trends in Propagation

Innovations in plant propagation have continually transformed the way we approach gardening and agriculture, with future trends promising even more groundbreaking advancements. The fusion of technology, biology, and environmental science is paving the way for more efficient, sustainable, and resilient propagation methods. For beginners and seasoned horticulturists alike, understanding these innovations can provide a glimpse into the future of plant cultivation and inspire new approaches to traditional practices.

One of the most exciting innovations in propagation is the use of automation and robotics. Automated systems are being developed to optimize every stage of the propagation process, from seed sowing to transplanting. These systems can perform repetitive tasks with precision, reducing labor costs and increasing productivity. Robotic arms equipped with sensors can select and plant seeds or cuttings with remarkable accuracy, ensuring uniformity and minimizing waste. As technology continues to advance, these automated systems are likely to become more accessible and affordable, making them a viable option for both large-scale commercial operations and smaller nurseries.

The integration of artificial intelligence (AI) and machine learning in propagation is also gaining momentum. AI algorithms can analyze vast amounts of data to optimize growing conditions, predict plant health issues, and recommend tailored interventions. For instance, AI-powered sensors can monitor environmental variables such as temperature, humidity, and light intensity in real-time, providing valuable insights into plant growth dynamics. These insights enable growers to make data-driven decisions, enhancing the efficiency and success rates of propagation efforts.

Vertical farming is another trend that is reshaping the landscape of plant propagation. By stacking layers of plants in controlled environments, vertical farms maximize space utilization and allow for year-round cultivation, regardless of external climate conditions. This method is particularly well-suited for urban areas, where space is limited but demand for fresh produce is high. Vertical farming systems often incorporate hydroponics or aeroponics, using minimal soil and water resources while maintaining optimal nutrient delivery. As urbanization continues to rise, vertical farming is poised to become a crucial component of sustainable urban agriculture.

The advent of synthetic biology is opening new frontiers in plant propagation. Scientists are exploring ways to engineer plants with enhanced traits, such as increased photosynthetic efficiency or improved resistance to environmental stressors. Synthetic biology techniques can also be used to develop novel plant-based materials and products, expanding the applications of propagation beyond food production. For example, researchers are investigating the potential of bioengineered plants to produce sustainable alternatives to plastic or to serve as biofactories for pharmaceuticals and other high-value compounds.

Gene editing technologies, such as CRISPR-Cas9, are revolutionizing the way we approach plant breeding and propagation. These tools allow for precise modifications to a plant's genetic makeup, enabling the development of new varieties with desirable characteristics. Unlike traditional breeding methods, which can be time-consuming and unpredictable, gene editing offers a rapid and targeted approach to achieving specific traits. This technology holds great promise for addressing global challenges such as food security and climate change, by creating crops that are more resilient to changing environmental conditions.

Sustainability is a driving force behind many of the innovations in propagation. As the global population continues to grow, there is an increasing need for efficient and environmentally friendly methods of plant production. Innovations in sustainable propagation include the use of biodegradable materials for growing media and containers, reducing reliance on non-renewable resources. Additionally, advancements in water-saving technologies, such as closed-loop irrigation systems and rainwater harvesting, contribute to more sustainable resource management.

The concept of "smart" greenhouses is gaining traction as a means of integrating various technological innovations into a cohesive system. These greenhouses are equipped with sensors, automation, and AI to create an intelligent growing environment that adapts to the needs of the plants. By continuously monitoring and adjusting environmental parameters, smart greenhouses can optimize plant growth and maximize yields while minimizing resource use. This holistic approach to propagation combines the best of technology and horticulture, offering a glimpse into the future of agricultural practices.

Looking ahead, the convergence of biotechnology, robotics, and data science in plant propagation holds immense potential for transforming the way we grow plants. These innovations not only promise to enhance productivity and efficiency but also offer solutions to some of the most pressing challenges facing agriculture today. From addressing food security to reducing environmental impact, the future of propagation is bright with possibilities.

For beginners interested in exploring these trends, starting small and staying informed is key. Experimenting with automation tools, vertical gardening techniques, or sustainable practices can provide valuable insights and inspire new ideas. Engaging with online communities, attending workshops, and following the latest research can help expand your knowledge and keep you abreast of emerging developments.

The journey into the future of propagation invites curiosity, creativity, and a willingness to embrace change. As we continue to innovate and explore new frontiers, the potential for growth and discovery is boundless. By staying open to new ideas and technologies, we can shape a future where plant propagation is not only more efficient and productive but also more sustainable and harmonious with the natural world.

Chapter 5: Tools and Materials for Propagation

Essential Equipment and Supplies

Navigating the world of plant propagation requires a thoughtful selection of equipment and supplies tailored to your specific needs and goals. Whether you're a beginner starting out with a small garden or an enthusiast looking to expand your propagation skills, having the right tools at your disposal can make all the difference. This chapter delves into the essential equipment and supplies needed for successful plant propagation, providing insights into their use and importance.

At the heart of any propagation endeavor lies the growing medium. The choice of medium can significantly impact the growth and health of your plants. Traditional options like seed-starting soil mixes are popular due to their balanced blend of nutrients and excellent drainage. However, for those interested in soilless methods, alternatives such as coconut coir, perlite, or vermiculite offer lightweight and sterile options that promote root development. Each medium has its unique properties, making it crucial to select one that aligns with the specific requirements of the plants you're propagating.

Containers play a vital role in plant propagation, providing a stable environment for seedlings and cuttings to develop. Seed trays and pots are commonly used for starting seeds, offering ample space for roots to grow. Biodegradable pots, made from materials like peat or paper, are an eco-friendly choice that can be planted directly into the ground, minimizing transplant shock. For those working with cuttings, propagation trays with individual cells or plug trays are ideal for keeping each cutting separate, reducing the risk of disease spread.

Humidity and moisture control are critical for successful propagation, as young plants and cuttings are particularly susceptible to drying out. A propagation dome or humidity dome is an invaluable tool for maintaining a consistent humidity level, creating a microclimate that encourages healthy root development. Pairing these domes with heat mats can further enhance the propagation environment by providing bottom heat, which is especially beneficial for germinating seeds or rooting cuttings that require warmth.

Lighting is another essential consideration, particularly for indoor propagation. Full-spectrum LED grow lights are highly recommended due to their energy efficiency and ability to mimic natural sunlight. These lights can be adjusted in intensity and duration to meet the specific needs of your plants, ensuring they receive the optimal amount of light for photosynthesis. Positioning lights at the correct height and angle can help prevent issues such as leggy seedlings, which occur when plants stretch towards inadequate light sources.

Proper tools are indispensable for efficient and effective propagation. A sharp pair of pruning shears or scissors is essential for taking clean, precise cuttings, minimizing damage to the parent plant and increasing the likelihood of successful rooting. A dibber, a small pointed tool, can aid in planting seeds or making holes for cuttings, ensuring they are placed at the correct depth. Additionally, a labeling system, whether through plant markers or tags, is vital for keeping track of different plant varieties and their propagation dates.

Watering equipment ensures your plants receive the right amount of moisture without causing waterlogging or nutrient leaching. A fine-mist spray bottle or watering can with a gentle rose attachment is ideal for delivering a light, even spray, preventing disturbance of the growing medium. Alternatively, self-watering systems can be a convenient option for maintaining consistent moisture levels, particularly in environments where manual watering may be challenging.

Fertilizers and nutrients are necessary to support the growth and development of propagating plants. While some growing media come pre-fertilized, additional nutrients may be required as plants develop. Liquid or water-soluble fertilizers allow for precise application, providing essential macro and micronutrients directly to the roots. It's important to follow recommended guidelines for concentration and frequency to avoid over-fertilization, which can harm young plants.

Pest and disease management tools are crucial for maintaining plant health throughout the propagation process. Sticky traps can help monitor and control flying insect populations, while horticultural oils and soaps offer a safe, natural way to manage pests. Maintaining cleanliness and hygiene in your propagation area is vital to prevent the spread of diseases. Regularly disinfecting tools, containers, and surfaces can significantly reduce the risk of contamination.

For those interested in exploring more advanced propagation techniques, additional equipment such as a misting system or aeroponics setup may be beneficial. These systems offer precise control over moisture and nutrient delivery, creating an optimal environment for delicate or high-value plants. While these setups may require a larger initial investment, they can offer significant benefits in terms of efficiency and success rates.

Embarking on a propagation journey equipped with the right tools and supplies sets the foundation for success. As you gain experience and confidence, you may find yourself expanding your collection of equipment and exploring new techniques. Each tool serves a purpose, contributing to the overall health and vitality of your plants. By understanding the role of each piece of equipment and how to use it effectively, you empower yourself to cultivate a thriving garden, full of diversity and beauty.

Choosing the Right Containers and Mediums

Selecting the appropriate containers and growing mediums is a pivotal step in successful plant propagation, directly influencing the growth, health, and productivity of your plants. Each plant species has unique requirements that dictate the type of container and medium best suited for its propagation. For beginners, understanding these nuances can transform a daunting task into a rewarding endeavor, laying the groundwork for a thriving garden.

Containers serve as the foundation for plant growth, providing support, structure, and protection for developing roots. The choice of container can affect everything from water retention to root development, making it essential to consider factors such as size, material, and drainage capabilities. Traditional plastic pots are widely used due to their affordability and availability, but they come in various shapes and sizes to accommodate different plant needs. When selecting a pot, it's important to ensure it has adequate drainage holes to prevent waterlogging, which can lead to root rot.

For those seeking eco-friendly alternatives, biodegradable pots made from peat, coir, or recycled materials are excellent choices. These containers gradually break down and integrate into the soil, reducing waste and minimizing transplant shock when moving plants outdoors. Fabric pots, another sustainable option, promote air pruning of roots, which encourages a healthier root system and prevents plants from becoming root-bound. The breathable nature of fabric pots also enhances aeration, supporting vigorous growth.

Container size is a critical consideration, as it influences the space available for roots to expand. Seedlings and small plants typically require smaller pots to maintain moisture and temperature stability, while larger, more established plants benefit from roomier containers that accommodate extensive root systems. Starting with the appropriate size can reduce the need for frequent repotting, minimizing stress on the plants and maintaining consistent growth conditions.

The choice of growing medium is equally important, as it provides the essential nutrients and support needed for plant development. Traditional potting soil is a popular choice for many gardeners, offering a balanced mix of organic matter and nutrients. However, for those interested in soilless propagation methods, options such as coconut coir, perlite, and vermiculite offer distinct advantages. Coconut coir is a renewable resource with excellent water retention and aeration properties, making it ideal for seedlings and cuttings.

Perlite, a lightweight volcanic glass, is commonly used to improve drainage and aeration in potting mixes. Its porous structure allows excess water to drain away while retaining some moisture, preventing overwatering and promoting healthy root growth. Vermiculite, on the other hand, is a mineral that expands when heated, enhancing its ability to retain water and nutrients. When used in combination with other mediums, vermiculite can improve moisture retention and nutrient availability, creating an optimal environment for plant propagation.

For those interested in hydroponic or aquaponic propagation, selecting the right inert medium is crucial. Expanded clay pellets, also known as LECA (lightweight expanded clay aggregate), are a popular choice for these systems due to their stable structure and ability to support plant roots without compacting. Rock wool, another common hydroponic medium, provides excellent support and moisture retention but requires careful handling to avoid skin irritation.

Understanding the specific needs of the plants you wish to propagate can guide your choice of containers and mediums. For example, succulents and cacti thrive in well-draining mediums like sandy soil or a mix of cactus potting mix and perlite, while ferns and tropical plants prefer moisture-retentive mediums rich in organic matter. Tailoring your approach to the unique requirements of each plant type can enhance their growth and resilience.

In addition to selecting the right containers and mediums, maintaining proper hygiene and care is essential for successful propagation. Sterilizing containers before use and ensuring mediums are free from pathogens can prevent the introduction of diseases that may harm your plants. Regularly checking for signs of pests or diseases and taking prompt action can protect your plants and promote a healthy growing environment.

As you gain experience and confidence in propagation, you may wish to experiment with different containers and mediums to discover what works best for your specific conditions and preferences. The world of propagation is vast and varied, offering endless opportunities for exploration and experimentation. By starting with the right tools and knowledge, you set the stage for a flourishing garden that reflects your dedication and passion for plant cultivation.

Embarking on the journey of choosing the right containers and mediums is both an art and a science, requiring a balance of practicality and creativity. Each decision you make contributes to the overall success of your propagation efforts, shaping the future of your garden. As you nurture your plants from seeds or cuttings to mature specimens, you'll develop a deeper appreciation for the intricate relationship between plants and their environment, enhancing your skills as a gardener and steward of the natural world.

The Importance of Sterilization and Hygiene

Maintaining sterilization and hygiene is a cornerstone of successful plant propagation, ensuring the health and vigor of your plants from the very start. Without proper attention to cleanliness, pathogens and pests can quickly take hold, undermining even the most well-planned propagation efforts. For beginners, establishing a routine of sterilization and hygiene not only protects plants but also sets the foundation for a thriving garden, free from the threats of disease and contamination.

In the world of plant propagation, the risks associated with neglecting hygiene are manifold. Pathogens such as fungi, bacteria, and viruses can rapidly spread in moist, warm environments, typical of propagation settings. These microorganisms, often invisible to the naked eye, can cause damping-off in seedlings, rot in cuttings, and a host of other problems that can be detrimental to plant health. By implementing effective sterilization practices, you can significantly reduce the risk of infection and promote robust growth.

The first step in maintaining a clean propagation environment is to sterilize all tools and equipment before use. Pruning shears, scissors, and knives should be disinfected between cuts, especially when moving from one plant to another. A simple solution of rubbing alcohol or a diluted bleach mixture can be used to wipe down tools, effectively eliminating potential pathogens. This practice is particularly important when taking cuttings, as the open wounds can serve as entry points for bacteria and fungi.

Containers, pots, and trays also require attention. New containers are generally clean, but reused ones should be thoroughly washed and disinfected before planting. Residual soil, organic matter, or water from previous uses can harbor diseases, which can infect new plants. A soak in a bleach solution followed by a rinse with clean water ensures that containers are free from harmful microorganisms. For those using biodegradable pots or materials, consider composting them after use to avoid recontamination.

The growing medium itself should be sterile, especially when starting seeds or rooting sensitive cuttings. Many commercial potting mixes undergo sterilization during production, but if you're preparing your own, it may be necessary to sterilize it at home. This can be done by heating the medium in an oven or microwave, effectively killing any pathogens present. While it may seem like an extra step, this precaution can prevent a host of issues down the line.

Work surfaces and propagation areas should be kept clean and organized. Regularly wiping down benches, tables, and shelves with disinfectant can reduce the buildup of dust, debris, and potential contaminants. Keeping the propagation area tidy not only minimizes disease risk but also creates an efficient workspace where tools and plants are easily accessible. An often-overlooked aspect of hygiene is personal cleanliness. Washing hands before handling plants or wearing disposable gloves can prevent the transfer of pathogens from your skin to the plants.

Water quality is another critical factor in maintaining plant health. Using clean, chemical-free water minimizes the introduction of pathogens and salts that can accumulate in the growing medium. If possible, collect rainwater or use distilled water for watering, as tap water may contain chlorine or other additives that can harm sensitive young plants. Additionally, ensure that watering cans, hoses, and other irrigation tools are kept clean and free from algae or mold.

Ventilation is vital in preventing the buildup of excess humidity, which can lead to the proliferation of fungal diseases. Good airflow helps maintain a stable microclimate, reducing the risk of mildew and other moisture-related issues. Fans can be used to circulate air, and opening windows or vents allows for natural air exchange. Ensuring that plants are not overcrowded also promotes better air circulation and reduces the risk of disease spread.

Integrated pest management (IPM) strategies can complement your sterilization and hygiene efforts. Regularly inspecting plants for signs of pests or disease allows for early intervention and control. Natural predators, such as ladybugs or predatory mites, can be introduced to manage pest populations, reducing the need for chemical treatments. If chemical controls are necessary, choose environmentally friendly options and apply them according to the manufacturer's instructions to minimize impact on the plants and surrounding environment.

Fostering a culture of cleanliness and vigilance in your propagation practices pays dividends in plant health and productivity. While it may require an investment of time and effort, the rewards are evident in the form of healthy, thriving plants that are more resilient to environmental stressors. As you establish these habits, they become second nature, seamlessly integrating into your propagation routine.

Understanding the importance of sterilization and hygiene in plant propagation is a lesson in the interconnectedness of all elements within the garden ecosystem. Each action you take, from cleaning tools to monitoring plant health, contributes to the overall success of your propagation efforts. This holistic approach not only safeguards your plants but also enhances your growth as a gardener, cultivating a deeper connection to the natural world and its cycles.

Setting Up a Propagation Workspace

Creating an ideal propagation workspace is akin to setting the stage for a symphony of growth, where every element harmonizes to support the thriving life of plants. For beginners, establishing such a space involves careful planning and thoughtful organization, ensuring that all necessary components are easily accessible and conducive to plant development. This chapter will guide you through the essentials of setting up a workspace that not only fosters healthy propagation but also enhances your experience as a gardener.

Location is the first consideration when establishing your propagation area. Ideally, the workspace should be in a well-lit area, receiving ample natural light to support the initial growth stages of young plants. A south-facing window or a sunroom can serve as an excellent spot, providing consistent light throughout the day. If natural light is limited, consider investing in full-spectrum grow lights to supplement and replicate the sun's rays, ensuring plants receive the necessary energy for photosynthesis.

Space planning within your propagation area is crucial. A clutter-free and organized workspace allows for efficient movement and reduces the risk of accidental damage to plants or equipment. Allocate distinct zones for different activities, such as seed sowing, potting, and watering. This segmentation helps streamline your workflow and maintains order, making your propagation efforts more enjoyable and less chaotic.

Surface selection is another important aspect of your workspace setup. Choose a sturdy, level table or bench that can withstand the weight of containers, tools, and other materials. The surface should be easy to clean and resistant to moisture, as spills and splashes are common occurrences during propagation. If possible, opt for a surface with an adjustable height to accommodate different tasks and reduce strain on your posture during extended gardening sessions.

Storage solutions play a vital role in maintaining an organized propagation area. Use shelves, cabinets, or pegboards to store tools, containers, and supplies neatly. Clear, labeled bins or jars can help keep seeds, fertilizers, and other small items easily accessible and prevent them from becoming lost or damaged. Ensuring that everything has its designated place minimizes clutter and allows you to focus on the task at hand without unnecessary distractions.

Water access is a practical consideration for your propagation workspace. Having a nearby water source, such as a sink or hose, can simplify the process of watering and cleaning. If this isn't feasible, consider using large-capacity watering cans or jugs to minimize trips to and from a distant water source. Ensuring that your watering tools are clean and well-maintained prevents the introduction of contaminants to your plants.

Ventilation is an often-overlooked aspect of a propagation workspace. Good airflow is essential for maintaining a healthy growing environment, as it helps regulate temperature and humidity levels while preventing the buildup of mold and mildew. If your workspace is indoors, consider using fans or opening windows to encourage air circulation. For outdoor setups, positioning plants in a way that allows for natural airflow can help maintain a stable microclimate.

Incorporating ergonomic considerations into your workspace design can make propagation a more comfortable and enjoyable experience. Use cushioned mats or supportive footwear to reduce fatigue during long periods of standing. Arrange frequently used tools and supplies within arm's reach to minimize bending and stretching. By prioritizing comfort, you create a space that invites you to spend more time nurturing your plants.

Safety precautions are essential in any workspace, including those dedicated to propagation. Ensure that your area is well-lit to prevent accidents and that all tools are stored safely when not in use. If you use chemicals or fertilizers, keep them in clearly labeled containers and out of reach of children or pets. Wearing protective gear, such as gloves or goggles, can safeguard you from potential hazards while working with plants and materials.

Creating a propagation workspace that reflects your personal style and preferences can enhance the overall gardening experience. Consider adding decorative elements, such as artwork or potted plants, to personalize your space and make it more inviting. A space that inspires you to return to it time and again becomes a sanctuary where you can unwind, recharge, and connect with nature.

As you set up your propagation workspace, remember that flexibility and adaptability are key. As your skills and interests evolve, so too might your needs and preferences for your workspace. Be open to making adjustments and improvements over time, whether it's upgrading equipment, reorganizing the layout, or experimenting with new techniques. A successful propagation workspace is one that grows with you, supporting your journey as a gardener and allowing you to continually refine your craft.

Establishing a dedicated propagation workspace is an investment in both your gardening endeavors and your personal well-being. By creating an environment that nurtures both plants and your own sense of satisfaction, you set the stage for a rewarding and fulfilling journey in plant propagation. As you spend time in this space, you'll witness the fruits of your labor take root and flourish, deepening your connection to the natural world and the art of nurturing life.

DIY Solutions for Propagation Enthusiasts

Embarking on the journey of plant propagation offers endless opportunities for creativity and innovation, particularly for those with a penchant for do-it-yourself (DIY) solutions. Crafting your own propagation tools and environments not only saves money but also allows for customization tailored to the unique needs of your plants. For enthusiasts eager to explore the art of propagation, this chapter provides a guide to inventive DIY solutions that cater to various aspects of the process, from tools to environments, enhancing both the practice and the joy of nurturing new growth.

One of the simplest yet most effective DIY projects is creating a propagation station using recycled materials. Glass jars, plastic bottles, or even old cups can be repurposed as containers for rooting cuttings. By filling these vessels with water or a suitable growing medium, you create a mini-environment where cuttings can develop roots before being transplanted. This approach not only reduces waste but also allows you to observe the fascinating process of root growth firsthand, providing both a functional and educational experience.

Humidity is a crucial factor in successful propagation, especially for cuttings that require a moist environment to prevent desiccation. Constructing a humidity dome from everyday household items can provide the necessary conditions for delicate plants. Clear plastic containers, plastic wrap, or even large zip-lock bags can serve as makeshift domes when placed over cuttings or seedlings. These DIY domes trap moisture and create a microclimate conducive to root development, ensuring that your plants remain hydrated and healthy.

For those interested in seed starting, a homemade seedling tray can be crafted from various recycled materials. Egg cartons, newspaper pots, or toilet paper rolls can be transformed into biodegradable seedling containers. These materials are not only cost-effective but also allow for easy transplanting, as the entire container can be placed directly into the soil, minimizing root disturbance. Additionally, using these materials contributes to sustainable gardening practices, reducing the reliance on plastic products.

Lighting is essential for the growth of young plants, and while commercial grow lights can be expensive, a DIY setup can be both effective and affordable. By repurposing a desk lamp or other adjustable light fixture, you can create a custom lighting solution. Pairing it with energy-efficient LED bulbs that emit the full spectrum of light necessary for photosynthesis allows you to provide adequate illumination for your plants. Adjusting the height and angle of the lamp ensures that your seedlings receive uniform light exposure, preventing issues such as legginess.

For those with a bit more technical prowess, constructing a DIY hydroponic or aquaponic system can be a rewarding endeavor. Using a basic setup of plastic containers, water pumps, and tubing, you can create a system that circulates nutrient-rich water to your plants. This method offers a soil-free alternative that can lead to faster growth rates and higher yields. By experimenting with different designs and setups, you can tailor the system to suit your specific plant varieties and growing conditions.

Creating a homemade rooting hormone can enhance the success rate of your cuttings. While commercial products are available, a natural alternative can be made using ingredients such as willow water or honey. Willow water, derived from soaking willow branches in water, contains natural rooting hormones that encourage root development. Honey, with its antiseptic properties, can protect cuttings from pathogens while promoting root growth. Applying these natural solutions to the cut ends of your cuttings can give them a head start in establishing roots.

Watering systems are another area where DIY ingenuity can shine. A simple drip irrigation system can be assembled using a plastic bottle with small holes poked into the cap. When filled with water and inverted into the soil, this makeshift system provides a slow, steady release of moisture, ensuring consistent hydration for your plants. This method is particularly useful for those who may not be able to water their plants frequently or for propagation areas that are difficult to access.

For propagation enthusiasts with limited space, vertical gardening solutions offer a practical and aesthetically pleasing option. By mounting shelves or repurposing pallets, you can create a vertical garden that maximizes growing space and adds a dynamic visual element to your propagation area. This approach is ideal for urban environments where horizontal space is at a premium, allowing you to cultivate a diverse array of plants in a compact setting.

Incorporating DIY solutions into your propagation practice not only fosters creativity but also deepens your connection to the process of plant cultivation. Each project offers an opportunity to experiment and learn, enhancing your skills and understanding of plant needs. The satisfaction of seeing your plants thrive in environments and systems you have crafted with your own hands is immeasurable, reinforcing the joy and fulfillment that comes from nurturing life.

As you explore these DIY solutions, remember that the key to successful propagation lies in adaptability and resourcefulness. Embrace the challenges and surprises that come with each project, as they are part of the journey toward becoming a more proficient and confident gardener. By infusing your propagation endeavors with creativity and innovation, you cultivate not only plants but also a lifelong passion for the art and science of gardening.

www.ingramcontent.com/pod-product-compliance
Lightning Source LLC
LaVergne TN
LVHW020332250125
802082LV00002B/293